The Golden Gift of Red Giants

Online at: https://doi.org/10.1088/2514-3433/adcf15

AAS Editor in Chief

Ethan Vishniac, Johns Hopkins University, Maryland, USA

About the program:

AAS-IOP Astronomy ebooks is the official book program of the American Astronomical Society (AAS) and aims to share in depth the most fascinating areas of astronomy, astrophysics, solar physics and planetary science. The program includes publications in the following topics:

GALAXIES AND
COSMOLOGY

INTERSTELLAR
MATTER AND THE
LOCAL UNIVERSE

STARS AND
STELLAR PHYSICS

EDUCATION,
OUTREACH,
AND HERITAGE

HIGH-ENERGY
PHENOMENA AND
FUNDAMENTAL
PHYSICS

THE SUN AND
THE HELIOSPHERE

THE SOLAR SYSTEM,
EXOPLANETS,
AND ASTROBIOLOGY

LABORATORY
ASTROPHYSICS,
INSTRUMENTATION,
SOFTWARE, AND DATA

Books in the program range in level from short introductory texts on fast-moving areas, graduate and upper-level undergraduate textbooks, research monographs, and practical handbooks.

For a complete list of published and forthcoming titles, please visit iopscience.org/books/aas.

About the American Astronomical Society

The American Astronomical Society (aas.org), established 1899, is the major organization of professional astronomers in North America. The membership (∼7,000) also includes physicists, mathematicians, geologists, engineers, and others whose research interests lie within the broad spectrum of subjects now comprising the contemporary astronomical sciences. The mission of the Society is to enhance and share humanity's scientific understanding of the universe.

The Golden Gift of Red Giants

Arlette Noels-Grotsch
Space Sciences, Technologies and Astrophysics Research (STAR) Institute, Université de Liège, Allée du Six-Août 19, 4000 Liège, Belgium

Andrea Miglio
Department of Physics & Astronomy "Augusto Righi", Alma Mater Studiorum – Università di Bologna, via Gobetti 93/2, 40129 Bologna, Italy

and

INAF-Astrophysics and Space Science Observatory of Bologna, via Gobetti 93/3, 40129 Bologna, Italy

and

School of Physics and Astronomy, University of Birmingham, Edgbaston B15 2TT, United Kingdom

IOP Publishing, Bristol, UK

ISBN   978-0-7503-2159-4 (ebook)
ISBN   978-0-7503-2157-0 (print)
ISBN   978-0-7503-2160-0 (myPrint)
ISBN   978-0-7503-2158-7 (mobi)

DOI   10.1088/2514-3433/adcf15

Multimedia content is available for this book from https://doi.org/10.1088/2514-3433/adcf15.

Version: 20250701

AAS–IOP Astronomy
ISSN 2514-3433 (online)
ISSN 2515-141X (print)

British Library Cataloguing-in-Publication Data: A catalogue record for this book is available from the British Library.

Published by IOP Publishing, wholly owned by The Institute of Physics, London

IOP Publishing, No.2 The Distillery, Glassfields, Avon Street, Bristol, BS2 0GR, UK

US Office: IOP Publishing, Inc., 190 North Independence Mall West, Suite 601, Philadelphia, PA 19106, USA

To Delphine and Catherine, to Emma and Anita

Contents

General Foreword   xii

Author biographies   xiii

1   Story of an Evolving Low Mass Star: Foreword   1-1

     References   1-2

2   Pre Main Sequence—PMS   2-1

2.1   Hayashi Track   2-1

     2.1.1   Fully Convective Structure   2-2

     2.1.2   Constant Effective Temperature   2-3

2.2   Mean Stellar Temperature during Gravitational Contraction   2-4

     2.2.1   Non-Degenerate Gas   2-4

     2.2.2   Degenerate Gas   2-5

2.3   Duration of the PMS Phase   2-7

2.4   List of Questions   2-7

     References   2-7

3   Core Hydrogen Burning   3-1

3.1   Minimum Mass for Nuclear Burning   3-1

3.2   Onset of Nuclear Reactions   3-2

     3.2.1   Thermostatic Pressure Control Mechanism (TPCM)   3-4

     3.2.2   Structural Adjustment toward Zero Age Main Sequence (ZAMS)   3-5

     3.2.3   Luminosity and Chemical Composition   3-6

3.3   Main Sequence (MS)   3-6

     3.3.1   Global Structural Changes during MS   3-6

     3.3.2   Main Sequence Lifetime   3-7

3.4   Terminal Age Main Sequence—Turn-Off   3-19

3.5   List of Questions   3-19

     References   3-20

4   Post Main Sequence   4-1

4.1   Formation of an Isothermal Helium Core   4-1

4.2   Schönberg–Chandrasekhar Mass Limit   4-1

4.3   Crossing the Hertzsprung Gap   4-5

| 4.4 | List of Questions | 4-8 |
| | References | 4-8 |

5   Red Giant Phase   **5-1**

5.1	Ages of Red Giants	5-1
5.2	Ascending the RGB	5-2
5.3	First Dredge-Up and RGB Bump	5-4
5.4	List of Questions	5-9
	References	5-9

6   Core Helium Burning   **6-1**

6.1	Onset of Core Helium Burning	6-1
	6.1.1 Low Mass Stars—$M \lesssim 1.8\ M_\odot$	6-1
	6.1.2 Intermediate Mass Stars—$M \gtrsim 2.4\ M_\odot$	6-5
	6.1.3 Transition Mass	6-6
	6.1.4 Red Clump and Secondary Clump	6-10
6.2	Quiescent Core Helium Burning	6-13
	6.2.1 Convective Core	6-13
	6.2.2 Induced Semiconvection	6-14
	6.2.3 Extra-Mixing	6-17
	6.2.4 Horizontal Branch	6-19
6.3	List of Questions	6-20
	References	6-20

7   Asymptotic Giant Branch   **7-1**

7.1	Early AGB Phase (EAGB Phase)	7-1
	7.1.1 Formation of a Helium Burning Shell (He-Shell)	7-1
	7.1.2 AGB Bump and Second Dredge-Up	7-1
	7.1.3 AGB Bump Constraint on Extra-Mixing	7-4
	7.1.4 Early ABG Phase	7-5
7.2	Thermally Pulsating AGB Phase (TP-AGB Phase)	7-6
	7.2.1 Pulse and Interpulse	7-6
	7.2.2 Third Dredge-Up	7-8
	7.2.3 S-Process Nucleosynthesis	7-9
	7.2.4 Luminosity at the Tip of the AGB	7-11
7.3	List of Questions	7-12
	References	7-13

8   On the Way to White Dwarf Cooling   **8-1**

8.1   List of Questions   8-2

9   Epilogue   **9-1**

10   Asteroseismology of Red Giant Stars   **10-1**

10.1 Global Oscillation Modes in Red Giant Stars   10-1

10.2 Inference on *Global Stellar Properties*   10-7

    10.2.1 Radii, Distances, and Masses   10-10

    10.2.2 Ages   10-11

    10.2.3 Mass Loss and Gain   10-13

    10.2.4 Investigating Internal Structures Using HRDs Enhanced   10-14
        with Asteroseismic Constraints

10.3 Direct Constraints on the *Internal Structure*   10-16

    10.3.1 Mixed-Mode Patterns: Average Period Spacing and   10-16
        Coupling between p- and g-Cavities

    10.3.2 Constraints on the *Internal Rotation Rate*   10-18

    10.3.3 Deviations from the Expected, Approximated, Frequency   10-18
        Patterns of p and g Modes

10.4 A Bright Future for Asteroseismology and Stellar Physics   10-19

    References   10-19

Bibliography   **11-1**

General Foreword

In the early days of asteroseismology, red giant stars were sort of left over targets with a very poor potential for bringing new and precise constraints to their structure and evolution. With previous and on-going space missions like CoRoT, *Kepler*, and TESS, the situation has drastically changed. Red giants are now among the best keystone targets of asteroseismology. Their masses, radii, ages, and evolutionary status can be determined with unprecedented precision. Moreover, the exquisite asteroseismic data are now supplemented with spectroscopic (APOGEE), and astrometric (Gaia) constraints, which offers a novel opportunity to recast and address long-standing questions concerning the evolution of stars and of the Galaxy.

For you, the reader, who plans to enter the vast domain of asteroseismology of red giants, we would like to show you the reasons of this turnaround as well as the richness of the information and constraints the red giants provide. We would like to show you *The Golden Gift of Red Giants*.

We begin by following the evolution of a low-mass star from its pre-main sequence phase to the final cooling of a white dwarf—a prerequisite for fully understanding how red giants differ from other stars and appreciating the wealth of information they can provide. We then devote the final section to exploring this wealth of information, as revealed by asteroseismology of red giants through detailed analyses of data from space-based telescopes.

Author biographies

Arlette Noels-Grotsch

Arlette Noels-Grotsch earned her PhD at the University of Liège in the "Stellar Evolution and Stellar Stability" group led by Paul Ledoux, a pioneering and prominent figure of the 20th century in the domain of theoretical astrophysics. Arlette's early research focused on the secular stability of stars, particularly stars undergoing a flash at the onset of nuclear reactions in degenerate stellar matter. Later on, she remained dedicated to stellar stability problems in the Sun, red giant stars, and massive stars. In this context, she served as the Belgian PI of the CoRoT space mission (2006–2014). During her whole career, she was especially committed to understanding all the ins and outs of stellar evolution. She presented her work in numerous astrophysics conferences and organized or co-organized several colloquia in Liège and abroad. She was happy to have the opportunity to share her knowledge with students and young researchers as she became a Professor of Theoretical Astrophysics at the University of Liège. She also had teaching responsibilities at the University of Louvain-la-Neuve and at the University of Namur.

Andrea Miglio

Andrea Miglio is Professor of Astrophysics at the University of Bologna. He obtained his PhD at the University of Liège and subsequently held academic positions at the University of Birmingham, UK. His research focuses on asteroseismology - the study of resonant oscillations in stars - as a tool to improve our understanding of stellar physics and to infer precise and accurate stellar properties. While his early work concentrated on the theoretical aspects of stellar structure and oscillations, he has increasingly focused on the direct comparison between models and data from space-based photometric missions. Alongside detailed studies of stellar interiors, he has also pursued the use of asteroseismic constraints to investigate the structure and evolution of the Milky Way. He has been closely involved in space missions such as CoRoT and *Kepler*, and is contributing to the scientific preparation of the upcoming ESA mission PLATO.

Arlette Noels-Grotsch and Andrea Miglio

Chapter 1

Story of an Evolving Low Mass Star: Foreword

We invite the reader to play the role of a detective and follow us investigating the structural changes encountered by a low or intermediate mass star during its life, from the pre-main sequence phase up to the final state of a cooling white dwarf, and trying to understand through questions, answers, and simple reasonings what are the physical phenomena responsible for these events.[1]

All the figures and shortened reasonings displayed in this textbook are available in attached pptx and Keynote files. The reader will also benefit from animated versions of some of the figures, describing the evolution of physical quantities in the course of time for each phase of stellar evolution. These animated figures are referenced in the text and are available in the online version of this book at http://doi.org/10.1088/2514-3433/adcf15.

Guidance Notes

1. Except when differently referenced, the results presented here for our guest star of 1.3 $M_\odot$ have been computed with the code CLES (Scuflaire et al. 2008) with $X = 0.71$ and $Z = 0.015$ and the AGSS09 metal abundances (Asplund et al. 2009), the FreeEOS equation of state (Irwin 2012), the reaction rates from Adelberger et al. (2011) for hydrogen burning and from Angulo et al. (1999) for helium burning, the OPAL opacities (Iglesias & Rogers 1996), with or without diffusion (Thoul et al. 1994), and with or without extra-mixing.

2. General notions about stellar evolution are assumed to be mastered. Stellar structure equations will of course be extensively used but won't be demonstrated.

3. In order to help us understand the physics underlying the evolution of our guest star, we shall extensively make use of "dimensional" or, more precisely, homology relations. They are obtained by assuming that stellar evolution

---

[1] For more "in depth" explanations, the reader will benefit from a thorough book learning, for instance in Kippenhahn & Weigert (1994), Salaris & Cassisi (2006), Maeder (2009), and Iben (2013a; 2013b). For problems typical of low mass stars, we also recommend the review by Catelan (2007). The advanced phases, especially the ABG phase, are extensively discussed in Lattanzio & Wood (2004).

doi:10.1088/2514-3433/adcf15ch1   1-1

proceeds through homology transformations, i.e., two consecutive models of masses M and M' and radii R and R' are such that each mass shells ($m/M = m'/M'$) are located at homologous points ($r/R = r'/R'$) (read, for instance, Chapter 20 in Kippenhahn & Weigert 1994). Depending on the specific context, they will be applied to the whole stellar structure or to strictly limited regions.

4. All the mass limits given in this volume are proxy values only, typical of standard models computed with a solar chemical composition. They slightly depend on various factors, such as chemical composition, for example, but also on the amount of extra-mixing during nuclear burning phases, as will be discussed at length. Better than symbols they allow a more comfortable magnitude scenery to help visualizing key points of physics at work in each phase of evolution of low and intermediate mass stars.

5. The list of references is not exhaustive at all. References have been chosen more to illustrate some point than to sort out the last word on the subject. They are thus a mixture of pioneering works and more advanced detailed and highly illustrative analyses.

6. The subjects presented in each chapter are discussed in the form of questions and answers. All the questions covered in a chapter are listed at the end of the chapter.

7. The text will closely follow the slides displayed in the enclosed keynote/pptx presentation entitled "Low Mass Stars."

References

Adelberger, E. G., García, A., Robertson, R. G. H., et al. 2011, RvMP, 83, 195

Angulo, C., Arnould, M., Rayet, M., et al. 1999, NuPhA, 656, 3

Asplund, M., Grevesse, N., Sauval, A. J., & Scott, P. 2009, ARA&A, 47, 481

Catelan, M. 2007, Graduate School in Astronomy: XI Special Courses at the National Observatory of Rio de Janeiro (XI CCE), ed. F. Roig, & D. Lopes, 39

Iben, I. 2013a, Stellar Evolution Physics, Volume 1: Physical Processes in Stellar Interiors (Cambridge: Cambridge Univ. Press)

Iben, I. 2013b, Stellar Evolution Physics, Volume 2: Advanced Evolution of Single Stars (Cambridge: Cambridge Univ. Press)

Iglesias, C. A., & Rogers, F. J. 1996, ApJ, 464, 943

Irwin, A. W. 2012, Astrophysics Source Code Library, record ascl:1211.002

Kippenhahn, R., & Weigert, A. 1994, Stellar Structure and Evolution (Berlin: Springer)

Lattanzio, J. C., & Wood, P. R. 2004, Evolution, Nucleosynthesis, and Pulsation of AGB Stars, Asymptotic Giant Branch Stars, Habing H. J. and Olofsson H. (eds) (Berlin: Springer) 23

Maeder, A. 2009, Physics, Formation and Evolution of Rotating Stars (Berlin: Springer)

Salaris, M., & Cassisi, S. 2006, Evolution of Stars and Stellar Populations (New York: Wiley)

Scuflaire, R., Théado, S., Montalbán, J., et al. 2008, Ap&SS, 316, 83

Thoul, A. A., Bahcall, J. N., & Loeb, A. 1994, ApJ, 421, 828

The Golden Gift of Red Giants

Arlette Noels-Grotsch and Andrea Miglio

Chapter 2

Pre Main Sequence—PMS

2.1 Hayashi Track

Why is there a forbidden region in the HR diagram?

On a bright galactic day an interstellar cloud is perturbed by a nearby supernova explosion and from its ensuing contraction, stars are formed. Among them a nice 1.3 $M_\odot$ star, our featured guest. Our story starts just after this rather violent phase, when the newborn star reaches hydrostatic equilibrium.

The temperature is low everywhere in the star and its only energy source comes from a global gravitational contraction. It is now located on the so-called Hayashi track in the Hertzsprung–Russell (HR) diagram, as can be seen in Figure 2.1.

Let us recall that according to the Schwarzschild criterion, convection occurs if

$$\nabla_{\rm rad} = \left(\frac{\partial \log T}{\partial \log P}\right)_{\rm rad} \geqslant \nabla_{\rm ad} = \frac{\Gamma_2 - 1}{\Gamma_2} \tag{2.1}$$

where rad and ad stand, respectively, for radiative and adiabatic, T and P are the temperature and the pressure, and Γ_2 is the second adiabatic coefficient. The equality marks the convective neutrality and fixes the boundary of a convective region. Most of the time, when this condition is satisfied, a value of the temperature gradient ($\nabla = \partial \log T / \partial \log P$) only very slightly larger than $\nabla_{\rm ad}$ is required to transport all the energy produced in the star. It is thus just assumed to be equal to $\nabla_{\rm ad}$ and the convection is said to be "adiabatic."

In his pioneering work, Hayashi (1961) showed that on the quasi-vertical track the temperature gradient is equal to $\nabla_{\rm ad}$ and the stellar structure is wholly convective while, in the shaded area to the right of the track, stars cannot reach hydrostatic equilibrium. If by accident, a star were located in this forbidden zone, its temperature gradient would be much greater than the adiabatic value and violent

doi:10.1088/2514-3433/adcf15ch2

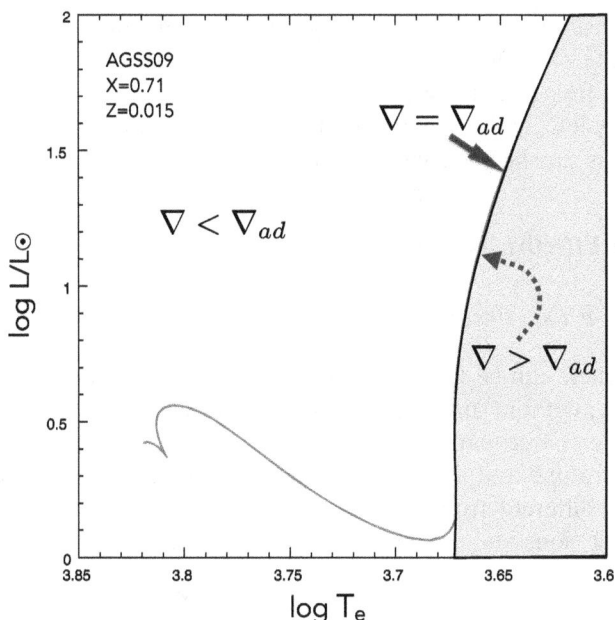

Figure 2.1. Quasi-vertical Hayashi track in the HR diagram for a 1.3 $M_\odot$ star. The shaded area is the forbidden region where hydrostatic equilibrium cannot be reached.

convective fluxes would hinder the establishment of hydrostatic equilibrium. However, those fluxes would have the effect of decreasing the temperature gradient and the star would land and stabilize on the Hayashi track. On the reverse, a star located to the left of the Hayashi track can be in hydrostatic equilibrium but must contain at least some radiative layers.

2.1.1 Fully Convective Structure

Are the physical conditions inside the star compatible with a fully convective structure on the Hayashi track?

The radiative temperature gradient can be written as

$$\nabla_{\rm rad} = \frac{3}{16\pi acG} \frac{P}{T^4} \frac{L}{m} \kappa \qquad (2.2)$$

where a, c, and G are, respectively, the radiation constant, the speed of light in vacuum, and the gravitational constant, and κ is the opacity coefficient. In the stellar interior where hydrogen is fully ionized κ closely follows a Kramers' law of opacity (see, for instance, Schwarzschild 1958)

$$\kappa = Cf(X, Y, Z)(1 + X)\rho T^{-3.5} \qquad (2.3)$$

where C is a constant, X and Y are the hydrogen and helium mass fractions, Z is the metallicity, and ρ is the density. The term $f(X, Y, Z)$ is equal to Z and $(X + Y)$ for bound-free and free–free transitions, respectively. In our newborn star the temperature is still very low, which means that the opacity is large. It is indeed so large that Equation (2.1) is satisfied everywhere and the star is wholly convective.

2.1.2 Constant Effective Temperature

Why is the effective temperature quasi-constant on the Hayashi track?

The photosphere can be seen to a first approximation as a region where a photon can travel freely, without any interaction with matter. Its thickness, h, is thus equal to a photon mean free path $(1/\kappa\rho)$ as illustrated in Figure 2.2, where T_e is the effective temperature and P_e is the associated pressure. In the photosphere the opacity is very different from Equation (2.3) since hydrogen is not ionized. It is dominated by H^- ions and can be crudely approximated by

$$\kappa_{H^-} \sim Z^{0.5}\rho^{0.5}T^9 \tag{2.4}$$

where the exponents of the density and the temperature slightly changes according to the physical conditions in the photosphere. The free electrons responsible for the formation of H^- ions are released from low ionization potential "electron donors" such as Si, Mg, Fe. Since the ionization of these donors is facilitated by a higher temperature the H^- opacity increases drastically with the temperature.[1]

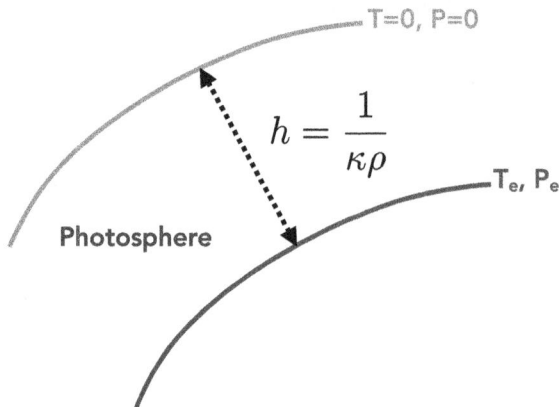

Figure 2.2. Schematic representation of the photosphere.

---

[1] This is true up to temperatures allowing hydrogen ionization.

Hydrostatic equilibrium at the base of the photosphere imposes that

$$P_e \sim \frac{GM}{R^2} \rho h \qquad (2.5)$$

where M and R are the total mass and radius of the star. If we assume the convection to be adiabatic below the photosphere, we have

$$T_e \sim P_e^{2/5}. \qquad (2.6)$$

From Equations (2.4), (2.5), and (2.6), we can write

$$T_e \sim \left(\frac{GM}{R^2} \frac{1}{\rho^{0.5} T_e^9} \right)^{2/5} \sim R^{-1/23} \qquad (2.7)$$

if ρ is dimensionally replaced by M/R^3. Equation (2.7) clearly shows that when contracting along the Hayashi track, a star keeps a quasi-constant temperature and a decreasing luminosity (since $L \sim R^2 T_e^4$).

2.2 Mean Stellar Temperature during Gravitational Contraction

How does the mean stellar temperature $\bar{T}$ behave as a result of contraction?

2.2.1 Non-Degenerate Gas

In the absence of external forces, the Viriel theorem writes

$$\Omega = \int_0^M -\frac{Gm}{r} \, dm = -3 \int_0^M \frac{P}{\rho} \, dm = -2U_t \qquad (2.8)$$

where Ω and U_t are, respectively, the total gravitational energy and the total internal energy. In a non-degenerate ideal gas with a negligible radiation pressure, the equation of state writes

$$P = \frac{k\rho T}{\mu m_H} \Rightarrow \frac{P}{\rho} \sim T \qquad (2.9)$$

where μ is the mean molecular weight $(2X + 3/4\,Y + Z/2)^{-1}$ and m_H is the atomic mass unit. If the star is contracting from a radius R_1 to a smaller radius R_2, we can write the variation of gravitational energy as follows:

$$\Delta\Omega = -2\Delta U_t \Rightarrow -GM^2 \left(\frac{1}{R_1} - \frac{1}{R_2} \right) \sim (\bar{T_1} - \bar{T_2}) \qquad (2.10)$$

and since R_1 is larger than R_2 it becomes clear that the mean temperature is increasing as a result of global contraction. Figure 2.3 shows the evolution of the central density ρ_c and central temperature T_c from the PMS to the Red Giant phase (see Chapter 5). The lower left part noted PMS displays the (ρ_c, T_c) progress during the fully convective part of the PMS contraction.

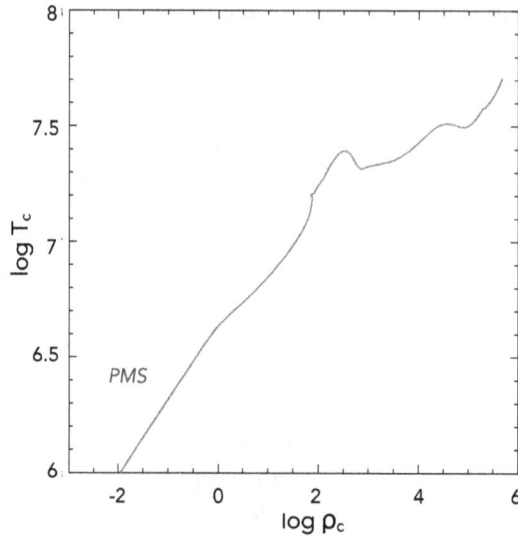

Figure 2.3. $\log T_c$ vs $\log \rho_c$ diagram for a 1.3 $M_\odot$ star. The lower left part noted PMS displays the fully convective part of the PMS contraction.

What happens if T increases?

Now from Equation (2.3), such an increase in temperature lowers the opacity, especially in the hottest central layers and the Schwarzschild criterion (Equation (2.1)) is no longer satisfied. A radiative core appears, the star becomes more transparent and the luminosity increases as can be seen in Figure 2.1 where the left turn is an indicator of the presence of a radiative core. The evolution of the central temperature and density during PMS can be seen in Figure 2.3 where a small change of slope signs the development of a radiative core in a so far fully convective structure.

2.2.2 Degenerate Gas

Is it always so?

However, this is not always the case. When the density becomes very large, the available space for each particle can become smaller than its "wave" volume λ^3, which writes

$$\lambda^3 = \left(\frac{h}{mv}\right)^3 \tag{2.11}$$

where λ is the de Broglie wavelength characterizing the spatial spread of the particle, h is the Planck constant, and m and v are the mass and the velocity of a particle, respectively. Introducing the following Maxwell–Boltzmann relation between the kinetic energy and the temperature, valid as long as the gas is non-degenerate

$$\frac{1}{2}mv^2 = \frac{3}{2}kT \qquad (2.12)$$

where k is the Boltzmann constant, we obtain the degeneracy criterion, which writes

$$\frac{\rho}{\mu}T^{-3/2} \geqslant \frac{(3km)^{3/2}}{h3}\,m_\mathrm{H}. \qquad (2.13)$$

It is clear from Equation (2.13) that the lightest particles are the first to become degenerate and the gas will thus consist of degenerate electrons and non-degenerate ions. The gas of electrons from now on will rely on Fermi–Dirac statistics and each elementary volume h^3 in the six-dimensional phase space (x, y, z, p_x, p_y, p_z) will host at most two electrons (see, for instance, Kippenhahn & Weigert 1994). At complete degeneracy all the elementary volumes of momentum smaller than a limit named the Fermi momentum are saturated. Any additional contraction leading to a higher density will require the filling of higher momentum elementary volumes, which results in a higher Fermi momentum. An increase of velocity is not anymore related to a higher temperature, as is the case in Maxwell–Boltzmann statistics, but is now related to a higher density. The equation of state for a completely degenerate gas reads

$$P = K\left(\frac{\rho}{\mu_\mathrm{e}}\right)^{5/3} \qquad (2.14)$$

where K is a constant and the exponent of (ρ/μ_e) turns to 4/3 if the degeneracy becomes relativistic. The internal energy is not proportional to the temperature anymore and the conclusion obtained with Equation (2.10) is not sustained.

The behavior of the mean temperature as a result of contraction can be obtained, assuming a homologous contraction, through the following dimensional argument. From hydrostatic equilibrium ($P \sim \frac{M^2}{R^4}$) and underlying meaning of density ($\rho \sim \frac{M}{R^3}$) non-degenerate, one can write

$$\frac{\delta P}{P} = \frac{4}{3}\frac{\delta\rho}{\rho}. \qquad (2.15)$$

Now, let us write the equation of state in a general form

$$P \sim \rho^\alpha T^\beta. \qquad (2.16)$$

It immediately follows that

$$\frac{\delta T}{T} = \left(\frac{4 - 3\alpha}{3\delta}\right)\frac{\delta\rho}{\rho}. \qquad (2.17)$$

In the non-degenerate case, we have

$$\alpha = 1 \text{ and } \delta = 1 \Rightarrow \frac{\delta T}{T} = \frac{1}{3}\frac{\delta\rho}{\rho} \qquad (2.18)$$

while in the degenerate case, this becomes

$$\alpha = \frac{5}{3} \text{ and } \delta = 0 \Rightarrow \frac{\delta T}{T} = -\infty\frac{\delta\rho}{\rho}. \qquad (2.19)$$

It follows from Equations (2.18) and (2.19) that a contracting star sees its mean temperature increase/decrease if the electron gas is non-degenerate/degenerate.

2.3 Duration of the PMS Phase

What is the duration of the PMS?

Th total gravitational potential energy of a star of mass M and radius R is given by (see Section 2.2)

$$|\Omega| = \frac{GM^2}{R} \qquad (2.20)$$

To estimate the time span of the PMS, we can assume that the star is radiating away all of its gravitational potential energy. The time it takes to do so at a luminosity L is thus given by

$$\Delta_{\text{PMS}} \sim \frac{GM^2}{RL} = t_{\text{KH}}, \qquad (2.21)$$

which is the Kelvin–Helmholtz timescale.

2.4 List of Questions

Why is there a forbidden region in the HR diagram?
Are the physical conditions inside the star compatible with a fully convective structure on the Hayashi track?
Why is the effective temperature quasi-constant on the Hayashi track?
How does the mean stellar temperature $\bar{T}$ behave as a result of contraction?
What happens if T increases?
Is it always so? What is the duration of the PMS?

References

Hayashi, C. 1961, PASJ, 13, 450
Kippenhahn, R., & Weigert, A. 1994, Stellar Structure and Evolution (Berlin: Springer)
Schwarzschild, R. 1958, Structure and Evolution of the Stars (Princeton, NJ: Princeton Univ. Press)

The Golden Gift of Red Giants

Arlette Noels-Grotsch and Andrea Miglio

Chapter 3

Core Hydrogen Burning

3.1 Minimum Mass for Nuclear Burning

What is the minimum mass for nuclear burning?

The impact of Relations (2.18) and (2.19) is extremely important for low-mass star evolution. This can easily be visualized in a $\log T_c$ versus $\log \rho_c$ diagram (Figure 3.1) similar to those that can be found in Kippenhahn & Weigert (1994). The most massive star M_3 can reach extremely high central temperatures while M_1 and M_2 only reach maximum central temperatures and become degenerate with cooler and cooler temperatures.

In order for a star to burn hydrogen in the core, the central temperature must be close to 10^7 K. Figure 3.1 shows that there exists a minimum mass for hydrogen burning (Hayashi & Nakano 1963; Nakano 2014), which is of the order of

$$T_c \sim 10^7 \text{K} \Rightarrow M_H \sim 0.08 M_\odot. \tag{3.1}$$

This mass is the minimum mass of a pure hydrogen star for which gravitational contraction is able to bring the central temperature to the onset value for hydrogen burning. A star whose mass is below the minimum mass for hydrogen burning is a *brown dwarf*. Its gravitational contraction increases its mean temperature up to a limit imposed by degeneracy and what follows is an irreversible cooling down.

Minimum masses are also encountered for nuclear burning of heavier elements. Helium burning requires a temperature of the order of 10^8 K while carbon burning starts around 5×10^8 K and we have

$$T_c \sim 10^8 \text{ K} \Rightarrow M_{He} \sim 0.33 \ M_\odot \tag{3.2}$$

for helium burning (Cox & Salpeter 1964), and

$$T_c \sim 5 \ 10^8 \text{ K} \Rightarrow M_C \sim 0.70 \ M_\odot \tag{3.3}$$

for carbon burning (Deinzer & Salpeter 1965).

doi:10.1088/2514-3433/adcf15ch3

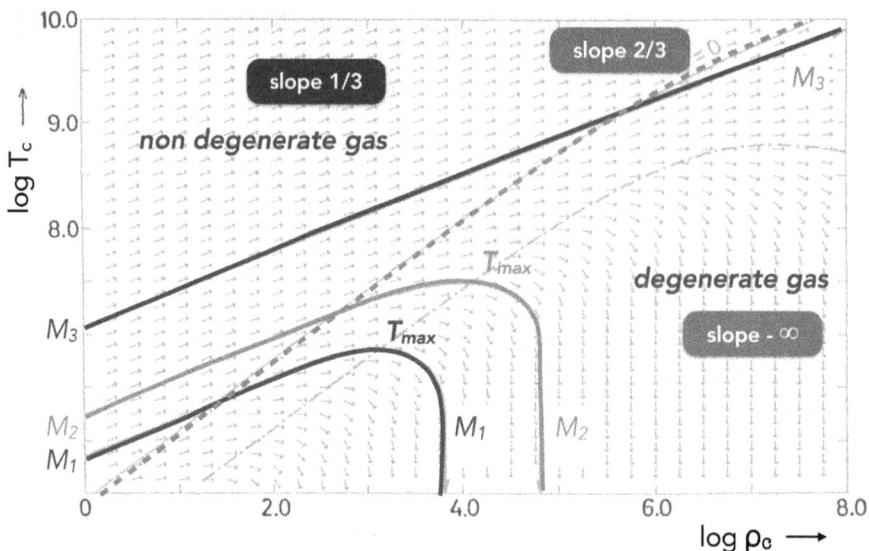

Figure 3.1. $\log T_c$ vs $\log \rho_c$ diagram. The dotted line displays the degeneracy criterion. Above/below that line, the equation of state is non-degenerate/degenerate. Red, green, and blue lines show the evolution of temperature and density in contracting chemically homogeneous stars of masses $M_1 < M_2 < M_3$. Reproduced from Kippenhahn & Weigert (1994), with permission from Springer Nature.

3.2 Onset of Nuclear Reactions

Fortunately our star of 1.3 $M_\odot$ is massive enough to start burning hydrogen. When its central temperature eventually reaches 10^7 K, nuclear reactions transforming hydrogen into helium take place. They are carried out either through the following proton-proton (pp) chain reactions or carbon-nitrogen-oxygen (CNO) cycle reactions.

pp Chain

$$
\begin{aligned}
{}^1_1\mathrm{H} + {}^1_1\mathrm{H} &\rightarrow {}^2_1\mathrm{H} + {}^0_1\mathrm{e} + \nu_e \\
{}^1_1\mathrm{H} + {}^2_1\mathrm{H} &\rightarrow {}^3_2\mathrm{He} + \gamma \\
\Rightarrow {}^3_2\mathrm{He} + {}^3_2\mathrm{He} &\rightarrow {}^4_2\mathrm{He} + 2{}^1_1\mathrm{H} \\
\Rightarrow {}^3_2\mathrm{He} + {}^4_2\mathrm{He} &\rightarrow {}^7_4\mathrm{Be} + \gamma \\
{}^7_4\mathrm{Be} + {}^0_{-1}\mathrm{e} &\rightarrow {}^7_3\mathrm{Li} + {}^0_0\nu \\
{}^7_3\mathrm{Li} + {}^1_1\mathrm{H} &\rightarrow {}^4_2\mathrm{He} + {}^4_2\mathrm{He} \\
\Rightarrow {}^3_2\mathrm{He} + {}^4_2\mathrm{He} &\rightarrow {}^7_4\mathrm{Be} + \gamma \\
{}^7_4\mathrm{Be} + {}^1_1\mathrm{H} &\rightarrow {}^8_5\mathrm{B} + \gamma \\
{}^8_5\mathrm{B} &\rightarrow {}^8_4\mathrm{Be} + {}^0_1\mathrm{e} + {}^0_0\nu \\
{}^8_4\mathrm{Be} &\rightarrow {}^4_2\mathrm{He} + {}^4_2\mathrm{He}
\end{aligned}
\tag{3.4}
$$

CNO Cycle

$$^{12}_{6}\text{C} + {}^{1}_{1}\text{H} \rightarrow {}^{13}_{7}\text{N} + \gamma$$
$$^{13}_{7}\text{N} \rightarrow {}^{13}_{6}\text{C} + {}^{0}_{1}\text{e} + {}^{0}_{0}\nu$$
$$^{13}_{6}\text{C} + {}^{1}_{1}\text{H} \rightarrow {}^{14}_{7}\text{N} + \gamma$$
$$^{14}_{7}\text{N} + {}^{1}_{1}\text{H} \rightarrow {}^{15}_{8}\text{O} + \gamma$$
$$^{15}_{8}\text{O} \rightarrow {}^{15}_{7}\text{N} + {}^{0}_{1}\text{e} + {}^{0}_{0}\nu \qquad (3.5)$$
$$^{15}_{7}\text{N} + {}^{1}_{1}\text{H} \rightarrow {}^{12}_{6}\text{C} + {}^{4}_{2}\text{He}$$
$$^{15}_{7}\text{N} + {}^{1}_{1}\text{H} \rightarrow {}^{16}_{8}\text{O} + \gamma$$
$$^{16}_{8}\text{O} + {}^{1}_{1}\text{H} \rightarrow {}^{17}_{9}\text{F} + \gamma$$
$$^{17}_{9}\text{F} \rightarrow {}^{17}_{8}\text{O} + {}^{0}_{1}\text{e} + {}^{0}_{0}\nu$$
$$^{17}_{8}\text{O} + {}^{1}_{1}\text{H} \rightarrow {}^{14}_{7}\text{N} + {}^{4}_{2}\text{He}$$

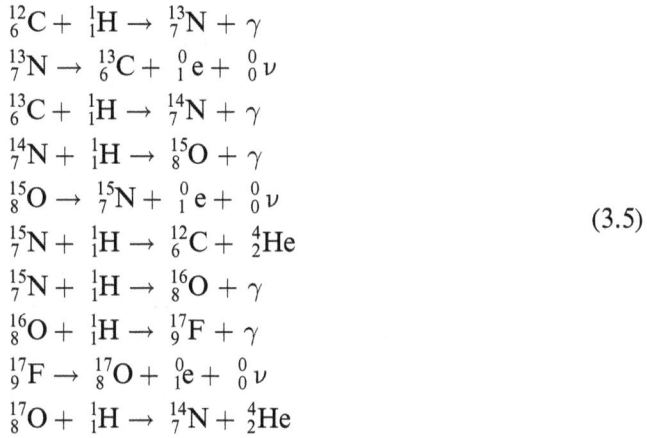

In the $\log T_c$ versus $\log \rho_c$ diagram displayed in Figure 3.2 the increase in T_c and ρ_c is momentarily stopped and there is an expansion and a cooling of the central layers (see inset in Figure 3.2).

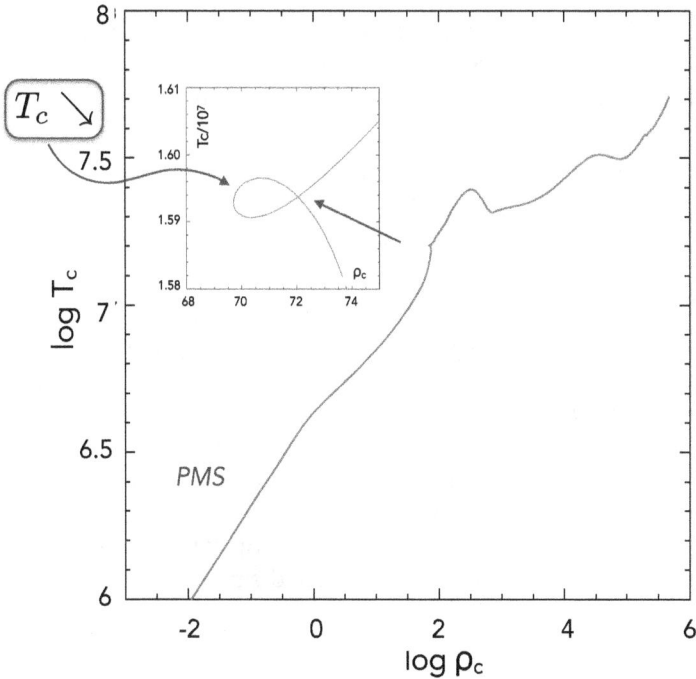

Figure 3.2. $\log T_c$ vs $\log \rho_c$ diagram for a 1.3 $M_\odot$ star. The inset is a zoom on the glitch related to the onset of nuclear burning near the center, showing the core expansion and the decrease in central temperature.

3.2.1 Thermostatic Pressure Control Mechanism (TPCM)

Why is there a cooling at the onset of nuclear reactions?

From the first law of thermodynamics and from Equation (2.16) we can write

$$dQ = dU + PdV = c_V\left(1 - \frac{3\delta(\Gamma_3 - 1)}{4 - 3\alpha}\right)dT = c^*dT \qquad (3.6)$$

where Q is the heat added to unit mass, U is its internal energy, V is its specific volume, c_V is the heat capacity at constant volume, and Γ_3 is the third adiabatic coefficient. Following Kippenhahn & Weigert (1994), we introduce the *gravothermal heat capacity* c^*.[1] In the non-degenerate case, we have

$$\alpha = 1 \text{ and } \delta = 1 \Rightarrow c^* = -c_V \qquad (3.7)$$

while in the degenerate case, this becomes

$$\alpha = \frac{5}{3} \text{ and } \delta = 0 \Rightarrow c^* = c_V. \qquad (3.8)$$

This means that adding heat to a non-degenerate gas induces a cooling while it produces a heating in a degenerate gas. The cooling of a non-degenerate gas at the onset of nuclear reactions may appear to be rather counterintuitive. It results from the conditions of hydrostatic equilibrium, which require that the central pressure P_c, which is given by the weight of the overlying layers ($P_{c,HE}$), be exactly equal to the pressure value related to T_c and ρ_c through the equation of state ($P_{c,EOS}$). This writes

$$P_c = P_{c,EOS} = \frac{k\rho_c T_c}{\mu M_H} = P_{c,HE} = \int_0^M \frac{Gm}{4\pi r^4} \, dm. \qquad (3.9)$$

If the central temperature T_c were to increase because of an added heat, an expansion would immediately act to prevent any increase in P_c, which would destroy the equality between $P_{c,EOS}$ and $P_{c,HE}$ (see Equation (3.9)). This is exactly what happens in stars more massive than M_H (see Equation (3.1)). This expansion leads in turn to a decrease in T_c.

This *thermostatic pressure control mechanism (TPCM)* acts as a thermostat, which prevents the gas from exceedingly heating up through a thermal runaway (see also Salaris & Cassisi 2006). Such a cooling of the central layers is then reversed by a global contraction and eventually the stellar structure adapts to its new energy source. This can be seen in Figure 2.1 as a loop with a decreasing luminosity resulting from the cooling, followed by a second smaller loop.

---

[1] The *gravothermal catastrophe* undergone by a system with a negative heat capacity has been described by Ledoux & Roger Wood (1968): Conductive transfer of heat from the central region will raise the high central temperature faster than it raises the lower temperature of the outer parts. No equilibrium is possible; the central layers continues to contract and gets hotter, sending out heat to the outer parts.

3.2.2 Structural Adjustment toward Zero Age Main Sequence (ZAMS)

How many loops before reaching the main sequence?

Stabilization on the main sequence (MS) means that all the reagents operating in the nuclear reactions have reached their "equilibrium" abundances, i.e., just as many nuclei of a specific isotope are formed and destroyed. This state is the Zero Age Main Sequence (ZAMS). The transformation of hydrogen into helium severely affects the internal profiles of all the reagents not only during the stabilization phase but also for the whole subsequent evolution as can be seen in Figure 3.3 illustrating the chemical profiles in our 1.3 $M_\odot$ now on the MS.

Figure 3.3 shows the profiles of X, He^3, C^{12}, N^{14}, and O^{16} in our 1.3 $M_\odot$ star at ZAMS (left panel) and later on at a more evolved MS state (right panel). In order to reach stabilization in pp chain, He^3 must first be accumulated (see Reaction (3.4)). This is responsible for the peak in He^3 that can be seen in both panels in Figure 3.3. To the right of the peak, He^3 has not reached its equilibrium value while to the left, the equilibrium value decreases as the temperature increases.

Low mass stars ($M \lesssim 1.1$ $M_\odot$) burn hydrogen essentially through pp chain and a single loop brings the star on the MS. More massive stars undergo hydrogen burning through CNO cycle reactions. In this cycle the reaction $N^{14}(p,\gamma)O^{15}$ (see Reaction (3.5)) has a cross section about 100 times smaller than the others. This means that a first loop occurs while accumulating N^{14}, which is followed by a second loop when achieving equilibrium for the full CN cycle (see Figure 2.1). This accumulation of N^{14} at the expense of C^{12} can be seen in both panels in Figure 3.3. At a more evolved MS state, with higher temperatures in the core, a reversal of N^{14} and O^{16} occurs when the full CNO cycle eventually operates (Figure 3.3, right panel).

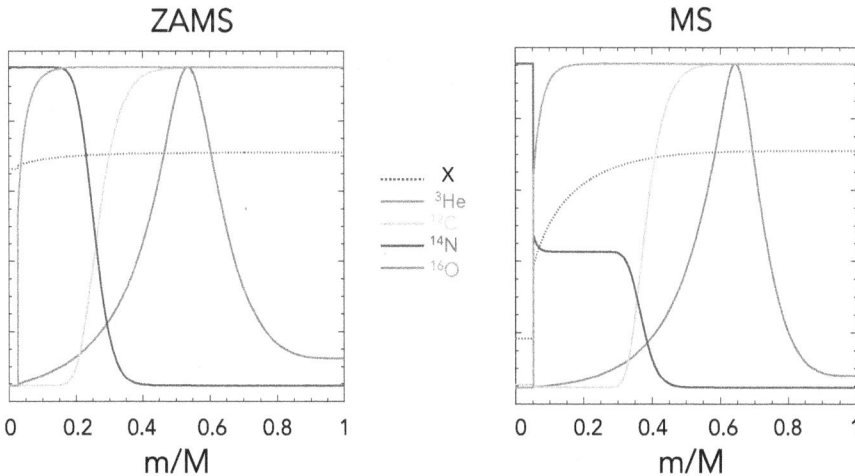

Figure 3.3. Profiles of He^3, C^{12}, N^{14}, O^{16}, and X in a ZAMS (left panel) and late MS (right panel) 1.3 $M_\odot$ star. Ordinate scales are the maximum values for each constituent except for hydrogen scaled to 1.

3.2.3 Luminosity and Chemical Composition

How does the ZAMS luminosity vary as a function of the chemical composition?

From the radiative transfer equation (Equation (2.2)) and the Kramers opacity law (Equation (2.3)), the luminosity can be written as

$$L = \nabla_{rad} \; m \; \frac{1}{\kappa} \frac{T^4}{P} \sim \frac{1}{Z(1 + X)\rho T^{-3.5}} \frac{T^4}{P}. \tag{3.10}$$

Adopting dimensional relations for $P\left(\frac{M^2}{R^4}\right)$, $\rho\left(\frac{M}{R^3}\right)$, and $T\left(\frac{M}{R}\right)$, one gets Schwarzschild (1958)

$$L \sim \frac{\mu^{7.5}}{Z}. \tag{3.11}$$

This very useful relation holds as long as the opacity obeys a Kramers-like form. For very massive stars with a scattering opacity ($\kappa \sim (1 + X)$) and a pressure dominated by the radiation pressure ($P \sim T^4$), the luminosity only very slightly varies with the hydrogen abundance ($L \sim (1 + X)^{-1}$).

3.3 Main Sequence (MS)

3.3.1 Global Structural Changes during MS

The rates of energy production through pp chain and CNO cycle operating at equilibrium can be approximated by the relations (see, for instance, Schwarzschild 1958)

$$\varepsilon_{pp} \sim C_{pp} X^2 \rho T^5 \quad \varepsilon_{CNO} \sim C_{CNO} X_{CNO} X \rho T^{15} \tag{3.12}$$

where C_{pp} and C_{CNO} are constant involving the energy released through the transformation of hydrogen into helium and X_{CNO} is the sum of the abundances of carbon, nitrogen, and oxygen. The temperature sensitivity in Equation (3.12) is slightly dependent on the physical conditions.

How does the luminosity, radius, and effective temperature vary during MS?

- **Luminosity:** During its evolution on the MS, our star slowly forms a helium-rich core, which affects the energy production rate since X decreases (see Equation (3.12)). This can however be compensated by a contraction of the core of mass m_n (where nuclear reactions take part) producing a slight increase in temperature. The heating of the core of mass m_n, radius r_n, and mean molecular weight μ_n, where nuclear reactions take part, can be understood by assuming a homologous contraction, which results in the following relations for the central pressure, central density, and central temperature[2]

$$P_c \sim \frac{m_n^2}{r_n^4}; \quad \rho_c \sim \frac{m_n}{r_n^3}; \quad T_c \sim \frac{P_c}{\rho_c}\mu_n \sim \frac{m_n}{r_n}\mu_n. \tag{3.13}$$

In a contracting core with an increasing mean molecular weight due to the formation of helium, T_c must thus increase. This temperature rise is however more important in the case of pp chain since its temperature sensitivity is smaller than in the case of CNO while X^2 intervenes instead of X (see Equation (3.12)). A side effect is a lowering of the opacity (Equation (2.3)) and the luminosity increases accordingly (Equation (2.2)). The luminosity increase is thus steeper for low-mass stars and L is nearly constant for higher mass stars burning with CNO.

- **Radius:** In order to keep the balance between the central pressure and the weight of the overlying layers, the contraction of the core (which could affect the weight of the overlying layers if it were a global contraction) must be compensated by an expansion of the hydrogen-rich envelope. The central pressure can indeed be written

$$P_c \sim \int_0^{m_n} \frac{Gm}{r^4} dm + \int_{m_n}^M \frac{Gm}{r^4} dm \tag{3.14}$$

where the radius r decreases with time in the first integral while it increases in the second integral. The overall result is a larger and larger total radius during MS.

- **Effective temperature:** In low-mass stars burning hydrogen mostly through pp chain, L and R both significantly increase and this results in a quasi-constant effective temperature. This is not the case for higher mass stars since L is nearly constant while R increases, which imposes a decrease in effective temperature.

This is illustrated in Figure 3.4. The transition from "mostly" pp chain to CNO cycle comes from the fact that the central temperature increases, which favors the more temperature dependent CNO cycle (see Relations (3.12)). Figure 3.5 shows the MS evolutionary tracks for a range of masses covering 0.9–1.5 $M_\odot$. The transition mass between stars burning hydrogen with pp chain and those burning with CNO cycle is of the order of 1.1 $M_\odot$ and slightly varies with the chemical composition.

3.3.2 Main Sequence Lifetime

What affects the core hydrogen burning lifetime?

---

[2] This expression for the temperature will also help us understand some physical behaviors in more evolved phases.

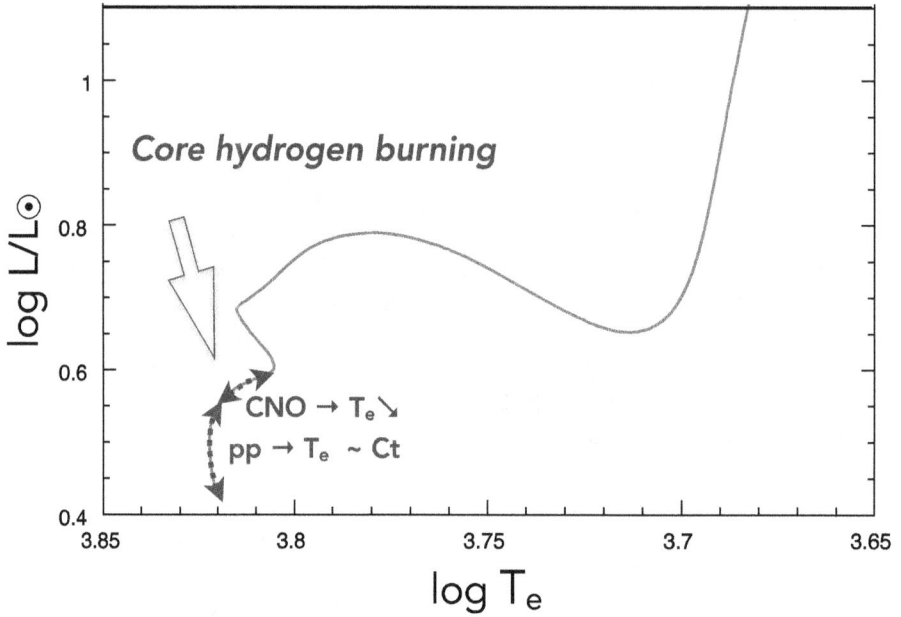

Figure 3.4. Core hydrogen burning in the Hertzsprung–Russell diagram for a 1.3 $M_\odot$ star.

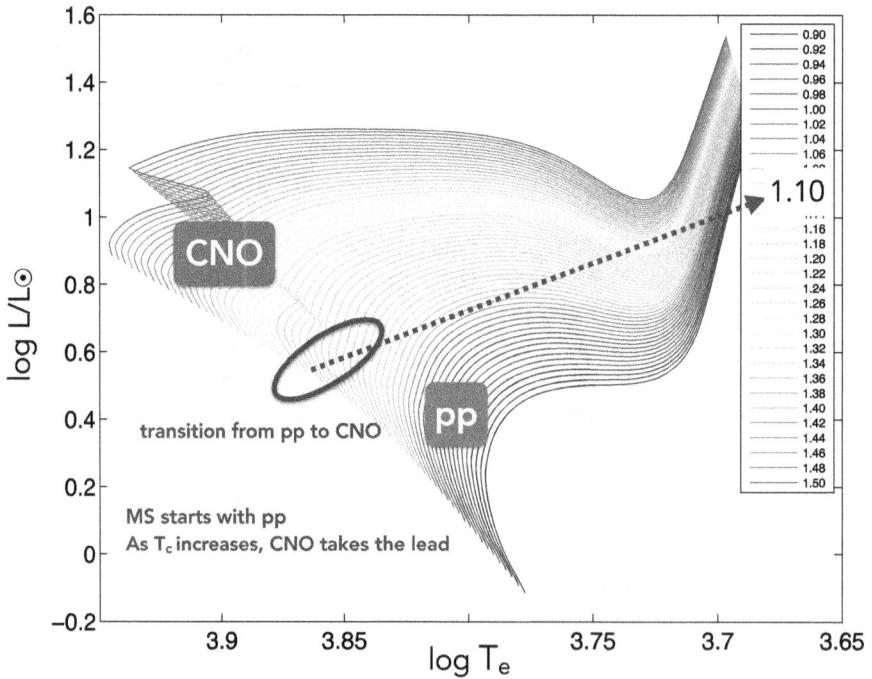

Figure 3.5. MS evolutionary tracks for a range of masses covering 0.9–1.5 $M_\odot$. Courtesy B. Rendle.

The MS lifetime, Δt_{MS}, may be estimated with the following relation:

$$\Delta t_{MS} \sim \frac{\Delta M_{H \to He}}{L} \tag{3.15}$$

where $\Delta M_{H \to He}$ is the amount of hydrogen transformed into helium during MS. This time span is much larger than the PMS lifetime, Δt_{PMS}. A 1 $M_\odot$ star for instance spends about 30 million years on the PMS versus 10 billion years on the MS.

Since the MS lifetime is inversely proportional to the luminosity, low metallicity stars have a shorter MS lifetime for a given X. However, the MS lifetime also depends on the amount of hydrogen burnt during MS. This brings us to the questions of the presence (or not) of a convective core, of the extent of mixing above stellar cores and of the impact of the chemical profile changes resulting from atomic diffusion on Δt_{MS}.

3.3.2.1 Convective Core

Is there a convective core during MS?

According to Equations (2.1) and (2.2), the occurrence of convection is favored not only by a high opacity but also by a large L/m ratio, which is equivalent to saying a highly concentrated energy source. Since we have, after reframing Equation (3.12),

$$L = \int_0^M \varepsilon_{pp/CNO} \, dm \sim \int_0^M \rho T^\nu \, dm \tag{3.16}$$

where the temperature sensitivity ν is of the order of 5 for the pp chain and 15 for CNO cycle reactions, it is obvious that the presence of the CNO cycle favors the presence of a convective core. However, when the pp chain operates out of equilibrium, i.e., when He^3 is still accumulating toward its equilibrium abundance, the temperature sensitivity of the pp chain is higher, of the order of 13, which also leads to the formation of a convective core.

The evolution of the extent of the convective core when He^3 is still out of equilibrium is shown in Figure 3.6. The temperature gradients $\nabla_{rad}, \nabla_{ad}, \nabla_T$ appear as full curves in red, cyan, and blue, respectively. The hydrogen abundance, X, and the helium abundance, Y are green and red dotted curves, respectively. As long as the radiative temperature gradient, ∇_{rad}, remains below the adiabatic gradient, ∇_{ad}, it is equal to the adopted temperature gradient, ∇_T, and the red curve is covered by the blue one. Convection occurs when $\nabla_{rad} > \nabla_{ad}$ and the red curve shows up while the blue curve now covers the cyan one since $\nabla_T = \nabla_{ad}$.

The evolution with central hydrogen abundance (X_c) of the convective core mass fraction (m_c/M) is illustrated in Figure 3.7 for a mass range covering 0.9–1.5 $M_\odot$. We can summarize it as follows:

- As long as He^3 has not reached its equilibrium abundance, the temperature sensitivity of ε_{pp} is large and a convective core is present. For masses smaller than about 1.1 $M_\odot$, this convective core disappears at the ZAMS.

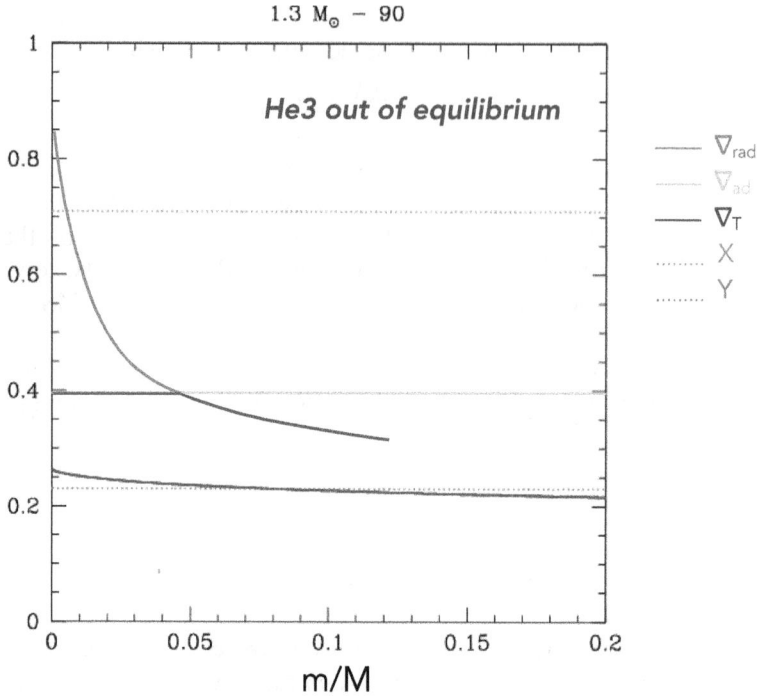

Figure 3.6. Figure showing the evolution of the convective core when He^3 is still out of equilibrium. The temperature gradients ∇_{rad}, ∇_{ad}, ∇_T are shown in red, cyan, and blue, respectively. The hydrogen abundance, X, and the helium abundance, Y, are displayed as green and red dotted curves, respectively. This figure corresponds to Movie1 in the supplementary Keynote and pptx files, accessible at http://doi.org/10.1088/2514-3433/adcf15.

- In core H-burning low-mass stars (1.1 $M_\odot \lesssim M \lesssim$ 2.0 $M_\odot$), the convective core mass grows with time during part of the MS as a result of (1) an out of equilibrium abundance of He^3 resulting from the fresh He^3 brought into the expanding convective core from the overlying layers, and (2) the larger and larger contribution of CNO cycle reactions due to the increasing temperature.
- In higher masses burning H with CNO cycle from the ZAMS, a convective core is present throughout MS and its mass decreases with time as a result of the decreasing opacity due to a lower and lower hydrogen abundance.

3.3.2.2 Semiconvective Region

Is there a semiconvective region during MS?

A problem encountered with a growing convective core mass is that of *semiconvection*, first discussed by Ledoux (1947). The larger and larger extent of the convective core progressively forms a discontinuity in chemical composition, which in turn introduces a discontinuity in the opacity and in the radiative temperature gradient (Equation (2.2)). This is illustrated in Figure 3.8, which shows that in some layers just outside the convective core, ∇_{rad} is larger than ∇_{ad}.

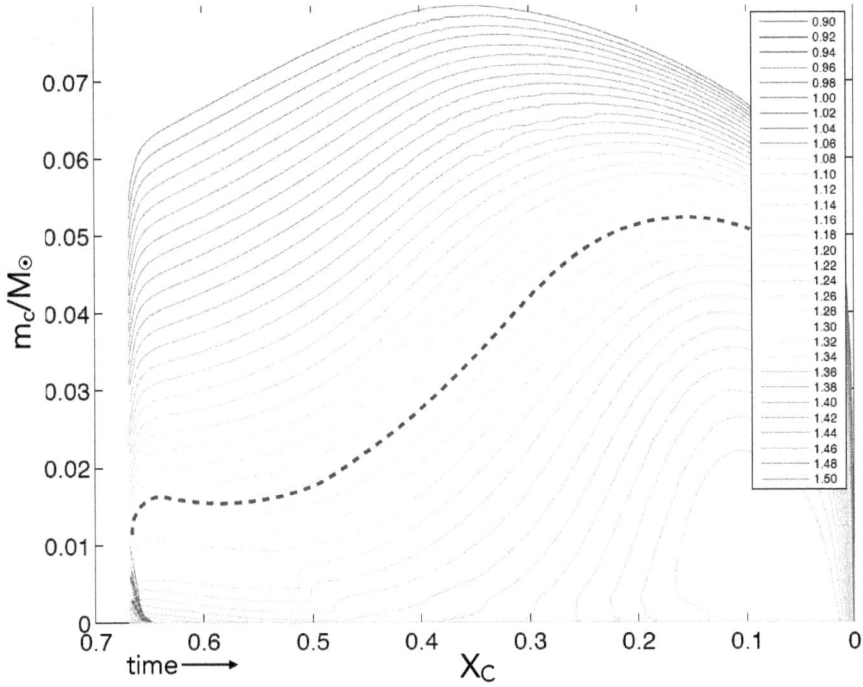

Figure 3.7. Evolution with central hydrogen abundance (X_c) of the convective core mass fraction (m_c/M) during MS for a range of masses covering 0.9–1.5 $M_\odot$. The dotted curve is the convective core mass fraction of a 1.3 $M_\odot$. Courtesy B. Rendle.

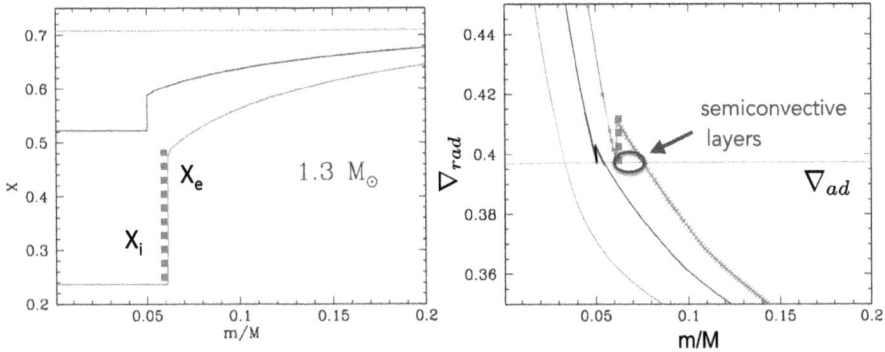

Figure 3.8. Illustration of the occurrence of a semiconvective region in a 1.3 $M_\odot$ star. The left panel shows the hydrogen profile in the inner 20% of the mass for three models on MSd, while the right panel displays $\nabla_{\rm rad}$ as a function of m/M for the same models. Reproduced with permission from Miglio et al. (2008).

The evolution of the extent of the convective core during MS is shown in Figure 3.9. The temperature gradients ∇_{rad}, ∇_{ad}, ∇_T appear as full curves in red, cyan and blue, respectively. The hydrogen abundance, X, and the helium abundance, Y are green and red the dotted curves, respectively. Due to the growing mass of the convective core, a semiconvective region appears. It is clearly visible as a peak in ∇_{rad} just at the outside of the convective core. It is also interesting to follow the changing profiles of X and Y during MS with a sharp discontinuity at the outer border of the convective core.

The transport of energy in such layers is still a matter of debate, especially as to the amount of mixing that can occur in the semiconvective region. For our 1.3 $M_\odot$ we have adopted the treatment recommended in Dziembowski (1977), i.e., an adiabatic temperature gradient and no mixing in the semiconvective layers.

The presence of a semiconvective region can often make it difficult to find the convective core boundary in stellar evolution computations and a special attention should be given to the correct use of the Schwarzschild criterion (Gabriel et al. 2014).

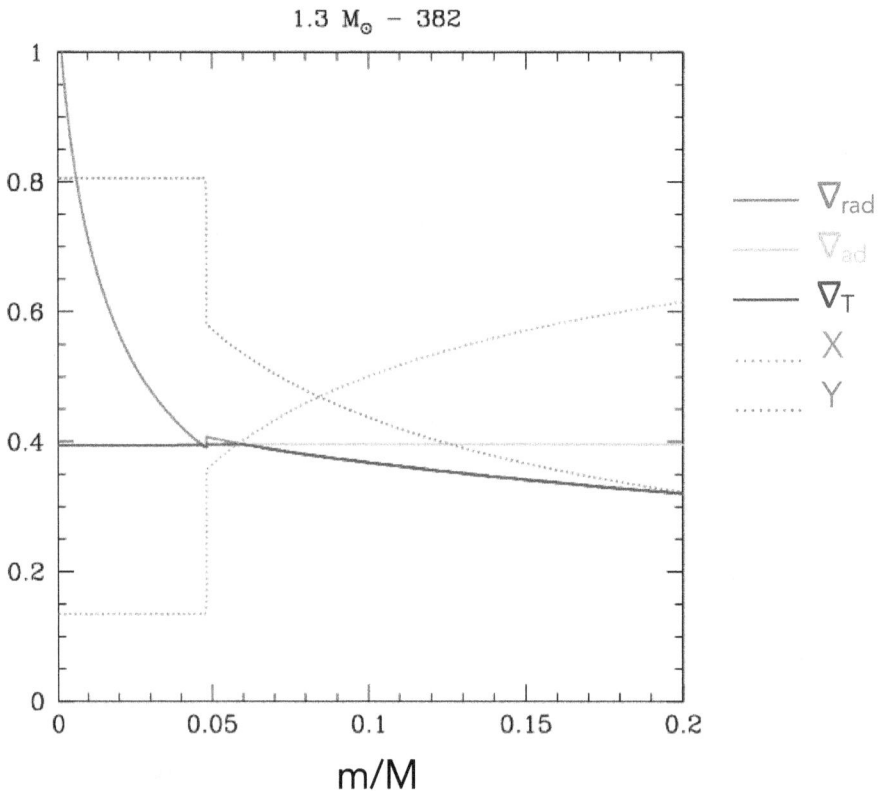

Figure 3.9. Figure showing the evolution of the convective core during MS. The temperature gradients ∇_{rad}, ∇_{ad}, ∇_T are shown in red, cyan and blue, respectively. The hydrogen abundance, X, and the helium abundance, Y are displayed as green and red dotted curves, respectively. This figure corresponds to Movie2 in the supplementary Keynote and pptx files, accessible at http://doi.org/10.1088/2514-3433/adcf15.

Otherwise the assumed boundary can be either the exact boundary or the external limit of the semiconvective region. This gives rise to an erroneous sawtooth behavior of the convective core boundary in the course of time.

3.3.2.3 Extra-Mixing

What are the impacts of adding an extra-mixing on top of the convective core?

The presence of an extra-mixing, either originating from convective bubbles overshooting the convective boundary or from rotational mixing seems to be well attested now through turn-off locations in clusters for example and even more precisely by fitting theoretical isochrones in double-lined eclipsing binaries (DLEBs) (Claret & Torres 2018 and references therein).

The way an extra-mixing is implemented in evolutionary codes should depend on its physical origin and is still a matter of debate. In most cases, an instantaneous mixing is assumed over a distance l_{ov} such that

$$l_{ov} = \alpha_{ov} H_P \qquad (3.17)$$

where H_P is the local pressure scale height and α_{ov} is an *overshooting parameter* generally smaller than unity.[3] If this distance is larger than the extent of the convective core, r_c, which often arises in low-mass stars, a fraction of r_c is preferred. The temperature gradient in the mixed region is also questionable, whether this mixing should result from a penetrative convection ($\nabla = \nabla_{ad}$) or a convective overshooting ($\nabla = \nabla_{rad}$) (see Zahn 1991). When the adiabatic gradient is adopted, the temperature gradient is larger than the radiative value, which is similar to a higher opacity. The luminosity is thus smaller and, as a result, the age is larger even though the burnt mass is smaller (see Noels et al. 2010 and references therein).

A *diffusive* mixing can also be adopted, for example, as a way to mimic rotational mixing (Ventura et al. 2008). Following Freytag et al. (1996) and Herwig et al. (1997), this mixing can be modeled through a diffusive coefficient, given by

$$D(r) = D_0 \exp\left(-2r/f_{ov} H_P\right) \qquad (3.18)$$

where r is the radial distance from the convective boundary, D_0 is the diffusion coefficient at the convective boundary, and f_{ov} is the diffusive overshooting parameter. Comparison of such models to evolutionary models computed with instantaneous mixing (Claret & Torres 2019) leads to a scaling factor of the order of

$$\frac{\alpha_{ov}}{f_{ov}} \sim 11.4. \qquad (3.19)$$

---

[3] Although the term *overshooting* is commonly used, let us not forget that besides its meaning of convective bubbles penetrating a radiative region, it may also be a way of describing in a simplified manner any other mixing such as rotational mixing.

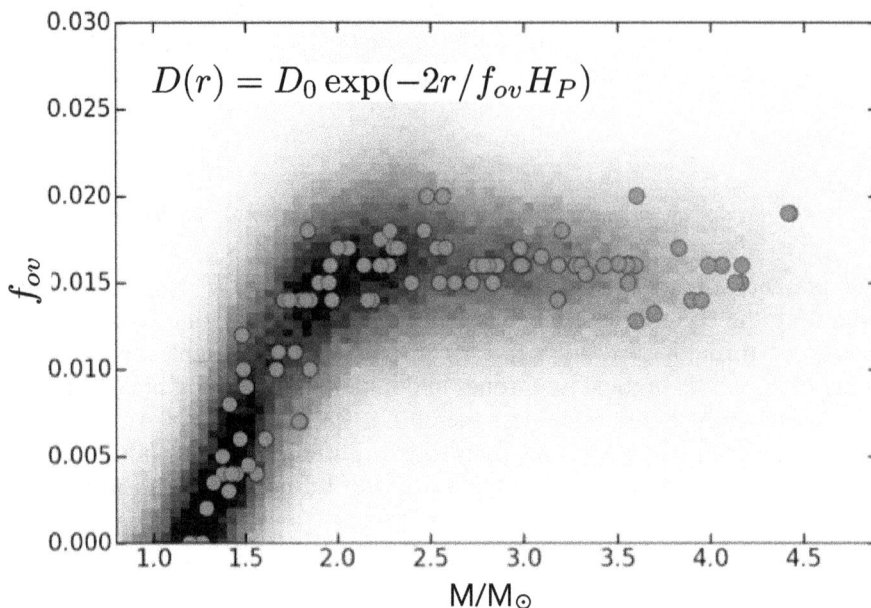

$$D(r) = D_0 \exp(-2r/f_{ov}H_P)$$

Figure 3.10. Evolution of the diffusive overshooting parameter with the stellar mass derived from isochrone fitting of DLEBs. Reproduced from Claret & Torres (2019). © 2019. The American Astronomical Society. All rights reserved.

Figure 3.10 shows the evolution of the diffusive overshooting parameter, f_{ov}, with respect to the stellar mass determined from isochrone fitting of DLEBs (Claret & Torres 2019). From ~1.1 $M_\odot$ to ~2 $M_\odot$, this parameter increases from 0 to ~0.018 (α_{ov} would range from 0 to ~0.2) and remains quasi-constant for higher masses. A somewhat similar behavior was obtained from seismic analyses of CoRoT and *Kepler* stars (Deheuvels et al. 2016). However, some words of caution have been advanced (see for instance Valle et al. 2016; Constantino & Baraffe 2018).

A "non-mixing length" theory, based on the plume theory, has been advanced to compute the temperature gradient in a shrinking convective core and its overshooting zone (Gabriel & Belkacem 2018). The extent of the overshooting region is given by the deceleration rate of bubbles rising from the convective core. These authors also address the case of a growing convective core mass in a low-mass MS star.[4] The presence of a density discontinuity produced by the forward displacement of the convective boundary aims at creating a density difference between both sides of the discontinuity, which rapidly becomes much larger than the convective density fluctuations. If the penetration by the convective elements into the stable zone is not able to smooth out the discontinuity, they simply crash into the barrier. This would prevent any overshooting *stricto sensu* at least during the growing phase of the convective core during MS. An overshooting related to a fraction of H_P could then

---

[4] This situation of a convective core mass growing with time is also encountered in core helium burning stars.

only exist during the last part of MS, when the convective core recedes, due to a lower and lower hydrogen abundance as in CNO burning more massive stars.[5]

This problem could however be mitigated by the presence of a semiconvective region just above the chemical discontinuity (see Section 3.3.2.2). Semiconvection can be at the origin of a partial mixing, which would iron out the discontinuity and ease the overshoot of convective bubbles. Material would however penetrate layers affected by a chemical gradient, which is a major difference from overshooting in more massive stars with receding convective cores, in which material moves within chemically homogeneous layers.

More advanced 3D analyses are performed to better represent mixing at and near convective boundaries, with an aim at integrating the results of 3D hydrodynamics simulations into a theoretical framework that can be used in 1D stellar evolution codes (Meakin & Arnett 2007, 2010; Cristini et al. 2019).

Adding some mixing on top of the convective core obviously increases the MS lifetime since more hydrogen is available for the formation of a helium core. This can be seen in the left panel of Figure 3.11 where the evolution with X_c of the convective core mass m_c/M and the mass of the whole mixed region m_{mix}/M are displayed for two values of the overshooting parameter, $\alpha_{ov} = 0.2$ and $\alpha_{ov} = 0.3$. In both cases, an instantaneous mixing has been adopted.

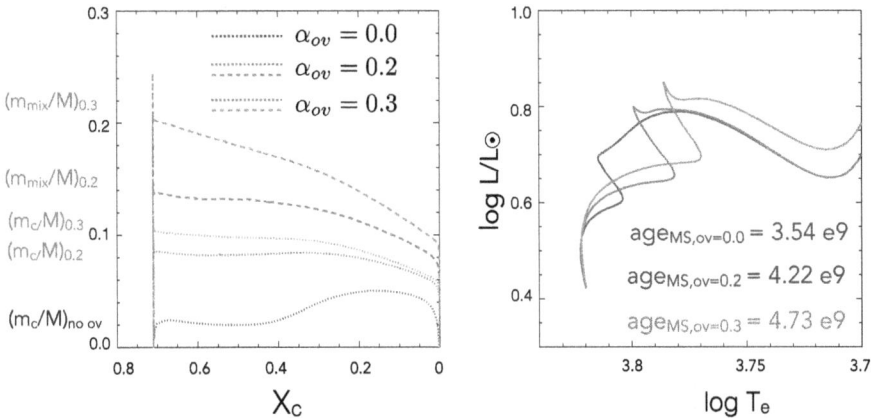

Figure 3.11. Left panel: Evolution with X_c of the convective core fractional mass (m_c/M, dotted line) and of the mixed region (m_{mix}/M, dashed line) in a 1.3 $M_\odot$ star for two values of the overshooting parameter, $\alpha_{ov} = 0.2$ (red) and $\alpha_{ov} = 0.3$ (green). The evolution of the convective core computed without any extra-mixing is drawn in blue. Right panel: Evolutionary tracks in the HR diagram of a 1.3 $M_\odot$ star for both values of the overshooting parameter (in red, $\alpha_{ov} = 0.2$; and in green, $\alpha_{ov} = 0.3$). For comparison the evolutionary track computed without any extra-mixing is drawn in blue. The MS lifetimes are indicated in the inserted legend.

---

[5] The theoretical models presented for our guest star have all been computed in a conservative way, with an overshooting distance proportional to H_P.

The right panel of Figure 3.11 illustrates the behavior of the evolutionary tracks in the HR diagram (in red and in green) computed with both assumptions for the extent of the extra-mixing. The track computed without extra mixing is shown in blue for comparison. The MS lifetimes are given in the inserted legend. We shall come back to this Figure 3.11 in Section 4.2.

An important side effect of extra-mixing in low-mass stars is the persistency of a convective core during a large part of MS, or during the whole MS, in stars which evolve with a radiative core when no extra-mixing is involved. Figure 3.12 shows the evolution of the fractional convective core mass in a star of 1.0 $M_\odot$ computed with different assumptions as to the extent of extra-mixing. Without any extra-mixing, the convective core disappears quite rapidly as soon as He^3 has reached its equilibrium abundance (see Section 3.3.2.1). Adding a small amount of extra-mixing already increases the time required for He^3 to reach its equilibrium abundance but for larger amounts, the mixing of a much larger region prevents He^3 from ever reaching its equilibrium abundance before the onset of CNO reactions (see the curve for $\alpha_{ov,MS}$).

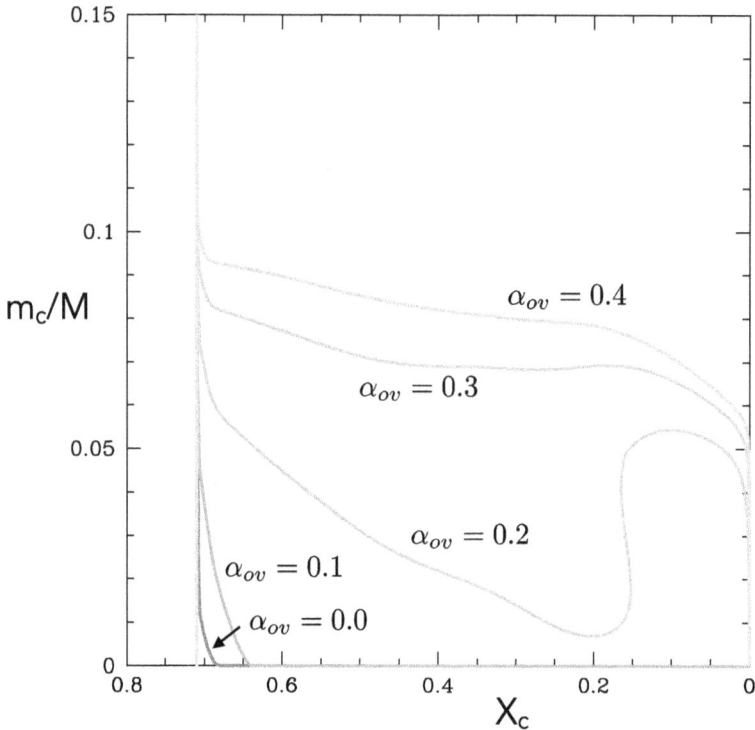

Figure 3.12. Evolution with X_c of the convective core fractional mass in a 1.0 $M_\odot$ computed with various assumptions of the extent of extra-mixing (listed in the legend).

A Word about Rotation

The impact of rotation and magnetic fields on stellar structure has been extensively discussed and analyzed by different groups, in particular by the Geneva group under the impulsion of A. Maeder and G. Meynet (see a very thorough reference book by Maeder 2009). The different physical aspects involved in the internal rotation of a star, as well as their implementation in the Geneva code, are fully described in Eggenberger et al. (2008). However, some important problems arise when comparing theoretical results with asteroseismic observations of low-mass stars (see, for instance, Eggenberger et al. 2012). They will be mentioned in Chapter 10.

3.3.2.4 Diffusion

Is diffusion at work in MS stars?

The importance of diffusion was definitely acknowledged when dating old globular clusters (GCs) (see Richard et al. 2002 and references therein). Without diffusion the ages of GCs were indeed larger than the age of the Universe. The main effect of diffusion is a gravitational settling, which deprives the surface layers of helium and heavy elements at the benefit of hydrogen while enriching the internal layers in helium and heavy elements. In GCs, MS stars are low-mass stars burning hydrogen through pp chain reactions with no convective cores. The gravitational settling of helium down to the center makes these stars "appear" more evolved than without diffusion, therefore reducing their MS lifetimes and the GCs ages through a *diffusion age effect*.

In the earlier days, diffusion had also very often be called for help in explaining chemical peculiarities in main sequence and hot horizontal branch stars (Section 6.2.4).[6] Moreover, one of the biggest successes of helioseismology has been the confirmation that microscopic diffusion was at work in the Sun Basu & Antia (1994).[7]

We have tested the impact of microscopic diffusion on the structure of our guest star. The left panel of Figure 3.13 shows the extent of the convective core in our 1.3 $M_\odot$ star computed without diffusion (blue dotted curve) and with diffusion (red dotted curve). The right panel of Figure 3.13 displays the HR evolutionary tracks for a 1.3 $M_\odot$ star computed with the same assumptions as in the left panel of Figure 3.13. The MS lifetimes are indicated in the figures.

The gravitational settling of helium in the core has an effect similar to the diffusion age effect found in GCs, i.e., a reduction of the MS lifetime. However, the difference is less important since our guest star has a convective core. Gravitational settling indeed increases the metallicity inside the core and enlarges the convective core. This can be seen in the left panel in Figure 3.13. Figure 3.14 displays the mass profile of the variables listed in the legend, for a model computed without diffusion

---

[6] For problems related to diffusion the reader will benefit from reading Vauclair (2003) and references therein.
[7] Another competing process is the levitation of chemicals by radiative accelerations. Very few codes have implemented this process (Théado et al. 2012) but, although it seems negligible in the Sun, it can have significant effects in more massive stars.

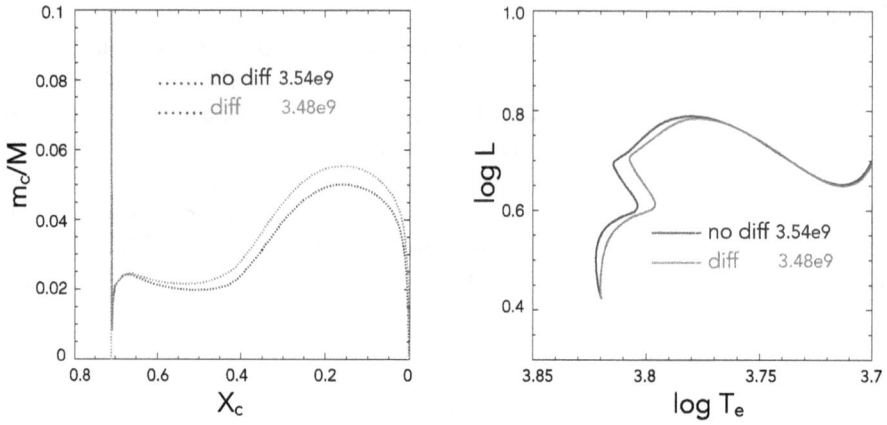

Figure 3.13. Left panel: Extent of the convective core in a MS 1.3 $M_\odot$ star computed without diffusion (blue dotted curve) and with diffusion (red dotted curve). Right panel: Evolutionary tracks of a 1.3 $M_\odot$ star computed without diffusion (blue curve) and with diffusion (red curve). The MS lifetimes are indicated in the figures.

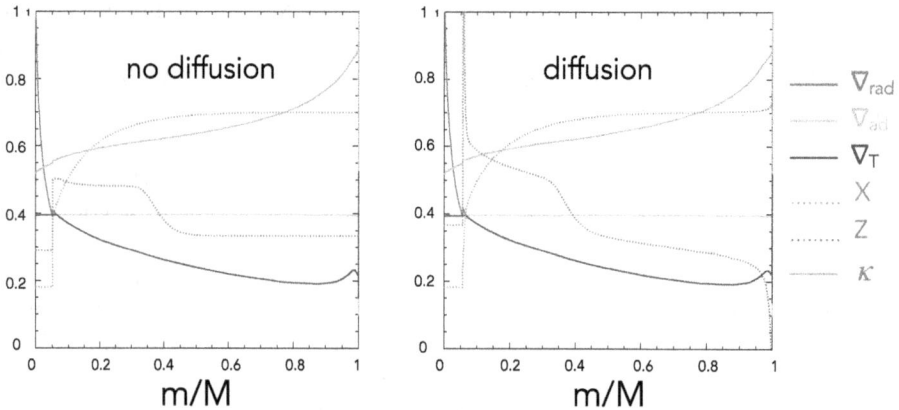

Figure 3.14. Comparison between the internal conditions listed in the legend in a model of an MS 1.3 $M_\odot$ star computed without diffusion (left panel) and a model computed with diffusion (right panel).

(left panel) and a model computed with diffusion (right panel). The increased metallicity throughout the star with the exception of the very external layers results in a lowering of the luminosity (see Equation (3.11)). Indeed, atomic diffusion slightly increases the size of the convective core while reducing the luminosity, both effects aiming at increasing the MS lifetime and thus reducing the diffusion age effect. It should also be noted that the accumulation of metals at the border of the convective core produces conditions even more favorable to semiconvection.

A Word about Radiative Accelerations
Although most stellar evolution codes include helium and heavy elements gravitational settling in the physical aspects of their models, albeit most of the time in a simplified way, very few add the effects of radiative accelerations (see Deal et al. 2016 and references therein). These accelerations produce depletion and/or accumulation of elements resulting from the competing gravitational, thermal, and radiative effects coupled with those due to the concentration gradients. This has a direct impact on the opacity, which is one of the main reasons why such computations are so complex and time consuming since the chemical mixture changes accordingly. In that respect the iron and nickel accumulations form the so-called "iron opacity bump" at $T \sim 2 \times 10^5$ K.

3.4 Terminal Age Main Sequence—Turn-Off

The exhaustion of hydrogen in the core marks the end of MS; this is the so-called terminal age main sequence (TAMS). In the HR diagram, this is seen as a *Turn-off* or a sharp bend toward higher effective temperatures, as shown in the left panel of Figure 3.11 and in Figure 3.5. Such a bend is however absent in the low-mass stars burning hydrogen essentially with pp chain reactions (Figure 3.5).

> *Why is there a bend at the TAMS of stars burning hydrogen with CNO cycle?*

We have seen that CNO cycle induces the presence of a convective core during MS while pp chain keeps the core radiative except at the onset of nuclear reactions when He^3 is still below its equilibrium abundance. It follows therefore that the exhaustion of hydrogen occurs in tiny central layers with pp chain while it affects a whole core in stars burning hydrogen with CNO cycle. Although it is a smooth process of moving the maximum of energy production in layers further off the center with pp chain, a somewhat drastic event occurs when CNO cycle is at work. This induces an overall contraction to counterbalance the shutdown of nuclear reactions in the core and the heating of the adjacent layers where hydrogen burning will take place from now on. Such a global contraction at a quasi-constant luminosity explains the increasing effective temperature observed during this phase also called "Second Gravitational Contraction."

Since the PMS phase is much shorter than the MS phase, the age at the turn-off, τ_{TO}, is of the order of the MS lifetime and we have

$$\tau_{TO} \sim \Delta t_{MS} \qquad (3.20)$$

3.5 List of Questions

What is the minimum mass for nuclear burning?
Why is there a cooling at the onset of nuclear reactions?
How many loops before reaching the main sequence?
How does the ZAMS luminosity vary as a function of the chemical composition?

How does the luminosity, radius and effective temperature vary during MS?
What affects the core hydrogen burning lifetime?
Is there a convective core during MS?
Is there a semiconvective region during MS?
What are the impacts of adding an extra-mixing on top of the convective core?
Is diffusion at work in MS stars?
Why is there a bend at the TAMS of stars burning hydrogen with CNO cycle?

References

Basu, S., & Antia, H. M. 1994, MNRAS, 269, 1137

Claret, A., & Torres, G. 2018, ApJ, 859, 100

Claret, A., & Torres, G. 2019, ApJ, 876, 134

Constantino, T., & Baraffe, I. 2018, A&A, 618, A177

Cox, J. P., & Salpeter, E. E. 1964, ApJ, 140, 485

Cristini, A., Hirschi, R., Meakin, C., et al. 2019, MNRAS, 484, 4645

Deal, M., Richard, O., & Vauclair, S. 2016, A&A, 589, A140

Deheuvels, S., Brandão, I., Silva Aguirre, V., et al. 2016, A&A, 589, A93

Deinzer, W., & Salpeter, E. E. 1965, ApJ, 142, 813

Dziembowski, W. 1977, AcA, 27, 95

Eggenberger, P., Meynet, G., Maeder, A., et al. 2008, Ap&SS, 316, 43

Eggenberger, P., Montalbán, J., & Miglio, A. 2012, A&A, 544, L4

Freytag, B., Ludwig, H. G., & Steffen, M. 1996, A&A, 313, 497

Gabriel, M., & Belkacem, K. 2018, A&A, 612, A21

Gabriel, M., Noels, A., Montalbán, J., & Miglio, A. 2014, A&A, 569, A63

Hayashi, C., & Nakano, T. 1963, PTP, 30, 460

Herwig, F., Bloecker, T., Schoenberner, D., & El Eid, M. 1997, A&A, 324, L81

Kippenhahn, R., & Weigert, A. 1994, Stellar Structure and Evolution (Berlin: Springer)

Ledoux, P. 1947, ApJ, 105, 305

Lynden-Bell, D., & Roger Wood, 1968, MNRAS, 138, 495

Maeder, A. 2009, Physics, Formation and Evolution of Rotating Stars (Berlin: Springer)

Meakin, C. A., & Arnett, D. 2007, ApJ, 667, 448

Meakin, C. A., & Arnett, W. D. 2010, Ap&SS, 328, 221

Miglio, A., Montalbán, J., Noels, A., & Eggenberger, P. 2008, MNRAS, 386, 1487

Nakano, T. 2014, 50 Years of Brown Dwarfs, ed. V. Joergens (Switzerland: Springer International) 5

Noels, A., Montalban, J., Miglio, A., Godart, M., & Ventura, P. 2010, Ap&SS, 328, 227

Richard, O., Michaud, G., Richer, J., et al. 2002, ApJ, 568, 979

Salaris, M., & Cassisi, S. 2006, Evolution of Stars and Stellar Populations (New York: Wiley)

Schwarzschild, M. 1958, Structure and Evolution of the Stars (Princeton, NJ: Princeton Univ. Press)

Théado, S., Alecian, G., LeBlanc, F., & Vauclair, S. 2012, A&A, 546, A100

Valle, G., Dell'Omodarme, M., Prada Moroni, P. G., & Degl'Innocenti, S. 2016, A&A, 587, A16

Vauclair, S. 2003, Ap&SS, 284, 205

Ventura, P., D'Antona, F., & Mazzitelli, I. 2008, Ap&SS, 316, 93

Zahn, J.-P. 1991, A&A, 252, 179

Arlette Noels-Grotsch and Andrea Miglio

Chapter 4

Post Main Sequence

4.1 Formation of an Isothermal Helium Core

Why is the helium core isothermal after the TAMS?

Without any energy production in the core, the temperature gradient vanishes and an isothermal core is formed. This core is surrounded by a hydrogen burning shell (H-shell) whose temperature is imposed by the fact that nuclear energy production in the shell must be compatible with the luminosity and should therefore remain quasi-constant. The building of an isothermal core results in the cooling of the central layers in order to form a plateau in the temperature profile at exactly the H-shell temperature. Figure 4.1 is once more a central temperature and density diagram for our favorite star. This cooling is clearly seen just after MS. The temperature plateau is visible in Figure 4.2.

This animated figure shows the temperature, density, and hydrogen profiles at the formation of an isothermal core. However, hydrogen is progressively depleted in the layers on top of the isothermal core, which means that the mass of this helium core, m_{is}, increases with time as indicated in Figure 4.2.

4.2 Schönberg–Chandrasekhar Mass Limit

Can the mass of the isothermal core grow forever?

Figure 4.3 schematically shows an isothermal core of temperature T, mass m_{is}, and radius r_{is} in a star of mass M and radius R. The mean molecular weight in the core is μ_{is}, while it is μ_{env} in the surrounding envelope. At the boundary of the helium core, the "inner" pressure is P_{is} while the "outer" pressure is P_{env}. Obviously, the condition for hydrostatic equilibrium is

doi:10.1088/2514-3433/adcf15ch4

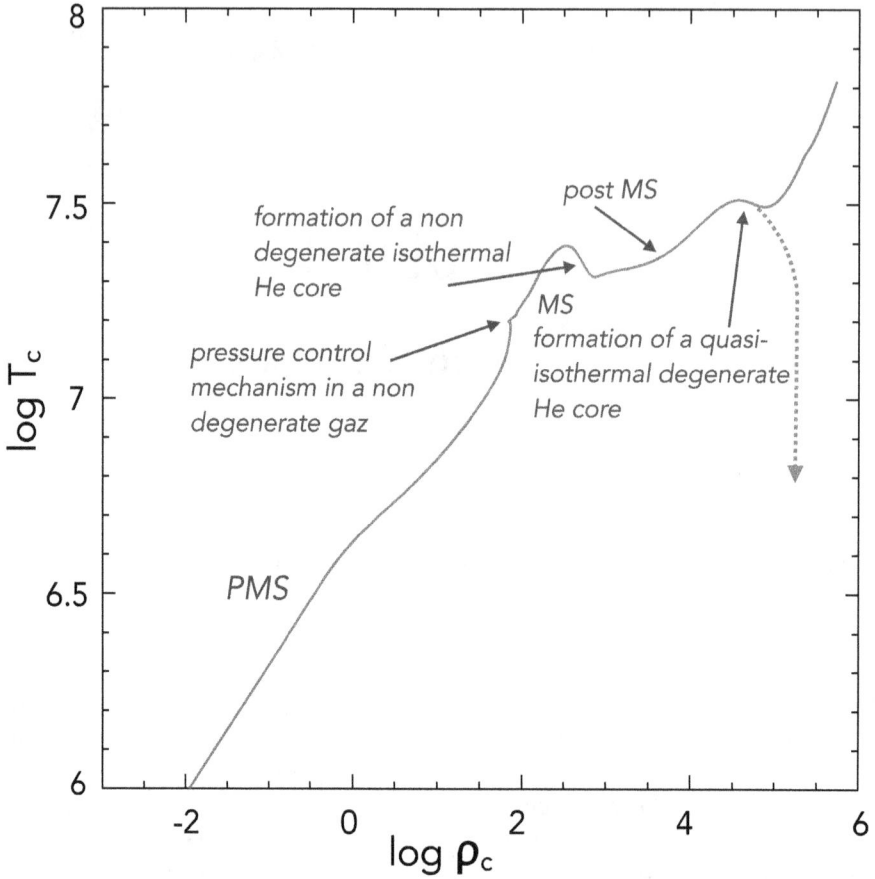

Figure 4.1. $\log T_c$ vs $\log \rho_c$ diagram for a 1.3 $M_\odot$ star. The cooling indicating the formation of an isothermal core is clearly seen after MS.

$$P_{is} = P_{env}. \tag{4.1}$$

Let us turn again to the Viriel theorem (Section 2.2). We can write

$$\Omega_{is} = -\int_0^{m_{is}} \frac{Gm}{r}\, dm = [P4\pi r^3]_0^{r_{is}} - 3\int_0^{m_{is}} \frac{P}{\rho}\, dm \tag{4.2}$$

$$= 4\pi r_{is}^3 P_{is} - 3\frac{kT}{\mu_{is} m_H} m_{is} \sim -\frac{G m_{is}^2}{r_{is}} \tag{4.3}$$

where the internal energy of the core has been expressed assuming that matter is non-degenerate in the isothermal core. It follows that

$$P_{is} \sim -3\,\frac{kT}{\mu_{is}}\frac{m_{is}}{4\pi r_{is}^3} - \frac{G m_{is}^2}{4\pi r_{is}^4}. \tag{4.4}$$

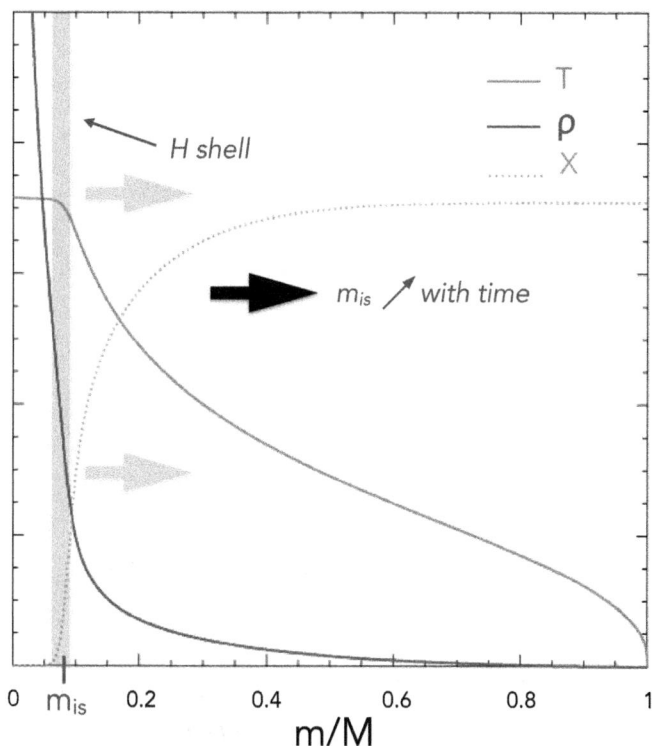

Figure 4.2. Figure showing the temperature, density, and X profiles in a 1.3 $M_\odot$ star after MS when an isothermal core of mass m_{is} is formed. This figure corresponds to Movie3 in the supplementary Keynote and pptx files, accessible at http://doi.org/10.1088/2514-3433/adcf15.

Equation (4.4) shows that for each value of m_{is} there is a maximum value of P_{is}, which is

$$\frac{\partial P_{is}}{\partial r_{is}} = 0 \Rightarrow P_{is,max} \sim \frac{T^4}{\mu_{is}^4 m_{is}^2} \tag{4.5}$$

and the maximum pressure $P_{is,max}$ decreases as m_{is} increases. Now P_{env} can be written as

$$P_{env} = \frac{k\rho T}{\mu M_H} = \int_0^M \frac{Gm}{4\pi r^4}\, dm \sim \frac{M^2}{R^4}. \tag{4.6}$$

The most massive isothermal core in the hydrostatic equilibrium frame is the one for which Equation (4.1) is fulfilled, which is in mass fraction

$$\frac{m_{is,max}}{M} = \frac{m_{SC}}{M} = \left(\frac{\mu_{env}}{\mu_{is}}\right)^2 0.37 \sim 0.08. \tag{4.7}$$

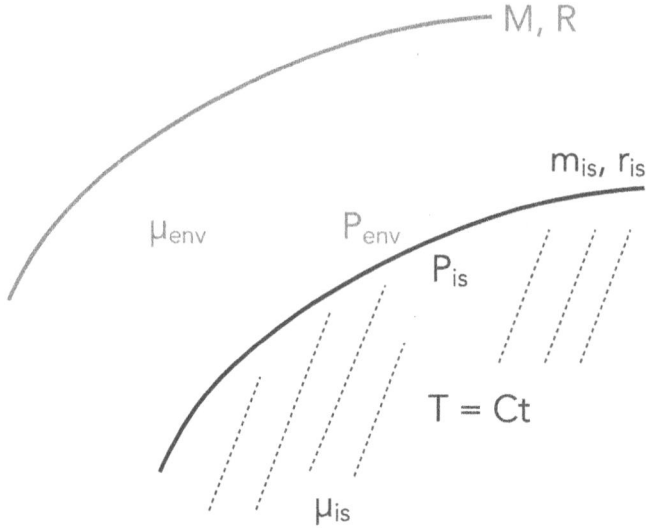

Figure 4.3. Schematic representation of an isothermal core of temperature T, mass m_{is}, and radius r_{is} in a star of mass M and radius R.

The Schönberg–Chandrasekhar mass limit m_{SC} (Schönberg & Chandrasekhar 1942) is the maximum mass allowed for an isothermal core in a star in hydrostatic equilibrium.

If the electrons in the core are degenerate, this notion of mass limit disappears since the internal energy of the core can now be written to a first approximation as

$$U_{t,is} = \frac{3}{2}\frac{kT}{\mu_{is}m_H}m_{is} + E_F\frac{m_{is}}{\mu_e m_H} \tag{4.8}$$

where E_F is the Fermi energy of the electrons. It can then be shown that $P_{is,max}$ becomes larger as m_{is} increases and there is no Schönberg–Chandrasekhar mass limit for an isothermal degenerate core.

Let us now return to the right panel of Figure 3.10. Without any extra-mixing our 1.3 $M_\odot$ star leaves MS with a helium core smaller than the Schönberg–Chandrasekhar mass limit. After the TAMS bend, the luminosity smoothly increases as the H-shell moves outward and the effective temperature slightly decreases as the envelope resumes its expansion (see the blue curve). The maximum in luminosity marks the Schönberg–Chandrasekhar limit for the isothermal core. With a large enough extra-mixing however, the Schönberg–Chandrasekhar mass limit is reached right after the TAMS and the slow increase in luminosity is replaced by a sharp zigzag pattern indicating the absence of an isothermal core. For our guest star, this happens for an overshooting parameter $\alpha_{ov,SC-TAMS} = 0.2$. If this parameter is smaller an intermediate situation occurs. If it is larger the helium core is already more massive than the Schönberg–Chandrasekhar mass limit at TAMS and this will keep an impact over the whole ulterior evolution.

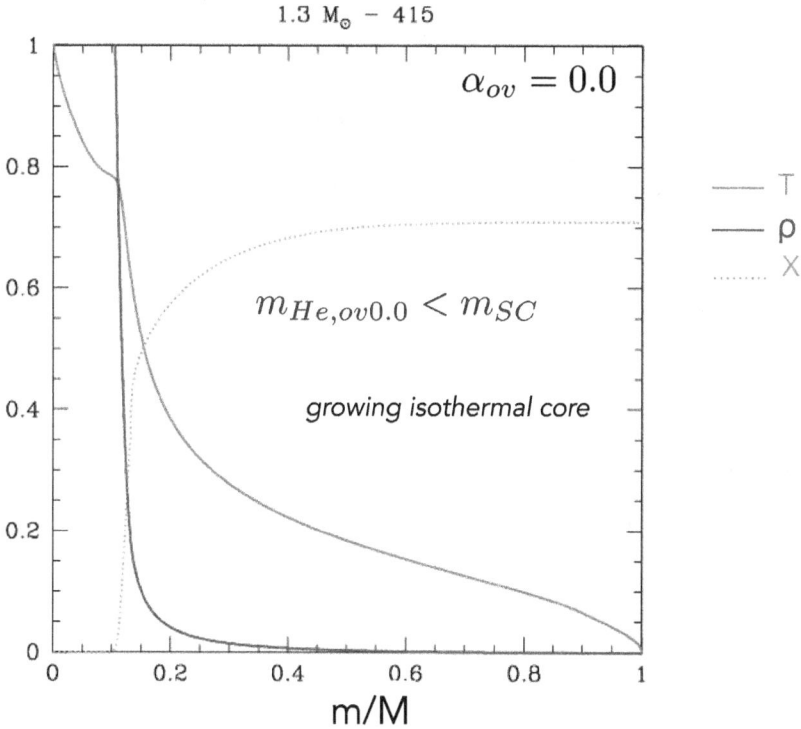

Figure 4.4. Figure showing the evolution of the temperature, density, and hydrogen profiles in the Post MS phase computed with an overshooting parameter $\alpha = 0.0$. This figure corresponds to Movie4 in the supplementary Keynote and pptx files, accessible at http://doi.org/10.1088/2514-3433/adcf15.

The evolution of the temperature, density, and hydrogen profiles computed with an overshooting parameter $\alpha = 0.0$ (no extra-mixing) are shown in Figure 4.4. Similar animations are available in the pptx and Keynote files for $\alpha = 0.2$ and 0.3.

Since, either at the maximum in luminosity, with no or small overshooting, or right after the TAMS, with an overshooting parameter of the order of $\alpha_{ov,SC\text{-}TAMS}$, the helium core mass is the same, roughly of the order of 0.1 M, the ages are similar and one has

$$\tau_{SC,ov} \approx \tau_{SC,no\ ov} \tag{4.9}$$

where τ_{SC} means the age at the Schönberg–Chandrasekhar mass limit.

4.3 Crossing the Hertzsprung Gap

After the helium isothermal core has reached the Schönberg–Chandrasekhar mass limit, our star must drastically adapt its structure to a new equilibrium configuration by rapidly contracting the core and expanding the envelope. The denser and denser core becomes degenerate while the cooler and cooler envelope sees its opacity increasing and convection sets in. The evolutionary track rapidly crosses the HR diagram toward the Hayashi track.

Why is the envelope rapidly expanding after having reached the Schönberg–Chandrasekhar mass limit of the isothermal core?

Figure 4.5 schematically shows a layer of mass Δm located in the envelope and above the H-shell, to which a perturbation δQ is applied in order to break thermal equilibrium while maintaining hydrostatic equilibrium. The evolution with time of such a perturbation underlies a *secular stability* analysis.[1] From Equation (3.6) and the equation of conservation of energy in the absence of nuclear reactions, we can write

$$\Delta \delta L = -\frac{\partial \delta Q}{\partial t} \, \Delta m = -\Delta m \, c^* \, \frac{\partial \delta T}{\partial t}. \tag{4.10}$$

From Equation (2.2) the luminosity is inversely proportional to the opacity, which temperature and density sensitivities are assumed to be κ_T and κ_ρ and we can write

$$L \sim \frac{1}{\kappa} \Rightarrow \delta L \ \sim -L\left(\kappa_T \frac{\delta T}{T} + \kappa_\rho \frac{\delta \rho}{\rho}\right) \sim -L\left(\kappa_T + \kappa_\rho \frac{3\delta}{4 - 3\alpha}\right)\frac{\delta T}{T} \tag{4.11}$$

and

$$\delta L_b - \delta L_t = \Delta m c^* \frac{\partial \delta T}{\partial t} \sim \ L((\kappa_{T,t} - 3\kappa_{\rho,t}) - (\kappa_{T,b} - 3\kappa_{\rho,b}))\frac{\delta T}{T} \tag{4.12}$$

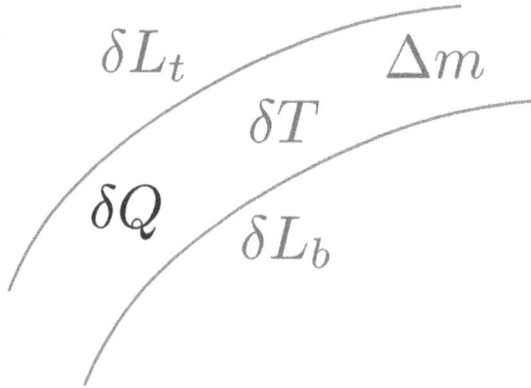

$$\delta L_t \qquad \Delta m$$
$$\delta T$$
$$\delta Q \qquad \delta L_b$$

Figure 4.5. Schematic representation of a layer of mass Δm in the envelope to which a perturbation δQ is applied to break thermal equilibrium while maintaining hydrostatic equilibrium. The perturbed temperature and luminosity are δT and δL. Indices b and t refer to the bottom and the top of the layer.

---

[1] Stars with an isothermal core mass smaller/larger than the Schönberg–Chandrasekhar mass limit are secularly stable/unstable.

since δ and α are equal to 1 in a non-degenerate matter with negligible radiation pressure and where the indices t and b refer to the top and the bottom of the layer.

As a result of the steeply decreasing temperature above the H-shell,[2] the temperature sensitivity is more negative on the cooler top of the layer while c^* is negative (Equation (3.7)). This means that $\delta T / T$ and $\partial \delta T / \partial t$ have the same sign. In other terms, a slight expansion of the layer leading to its cooling induces more and more cooling and expansion and a thermal runaway ensues. On the contrary, the core contracts and heats up, as can be seen in Figure 4.1.

This rapid crossing of the HR diagram is visible in Figure 4.6 as a lack of star in between the MS and the Red Giant Branch (RGB).

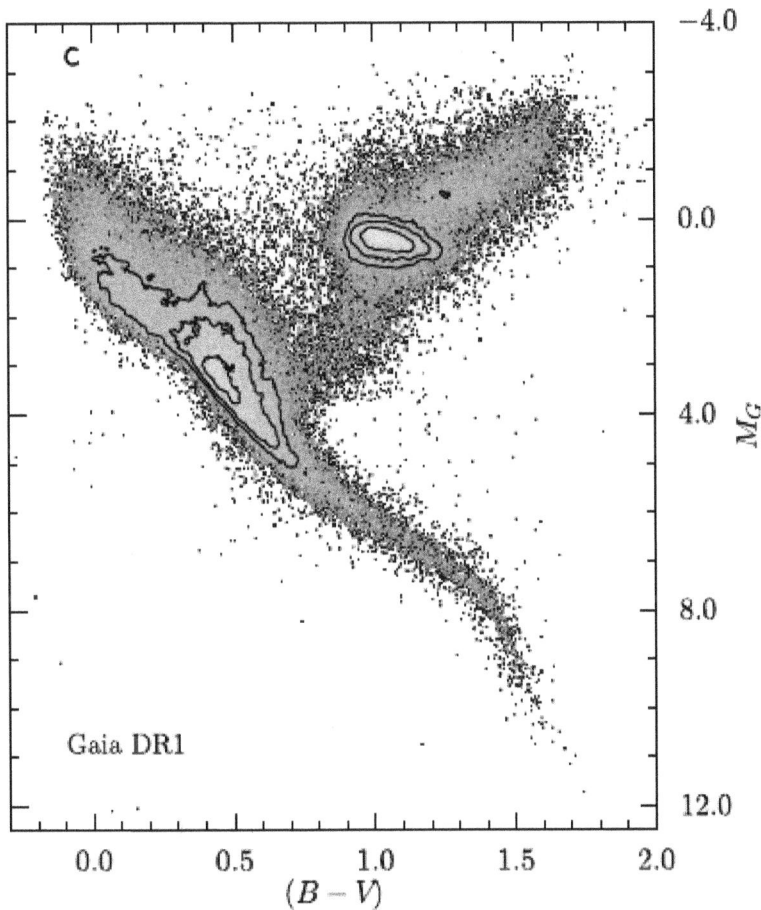

Figure 4.6. Hertzsprung–Russell diagram of DR1 stars observed by the Gaia mission. Credit: Gaia Collaboration et al. (2016), reproduced with permission © ESO

---

[2] The contribution of electron scattering is more important at the bottom of the layer where the temperature is significantly larger than at the top border.

4.4 List of Questions

Why is the helium core isothermal after the TAMS?
Can the mass of the isothermal core grow forever?
Why is the envelope rapidly expanding after having reached the Schönberg–Chandrasekhar mass limit of the isothermal core?

References

Gaia Collaboration, Brown, A. G. A., Vallenari, A., et al. 2016, A&A, 595, A2
Schönberg, M., & Chandrasekhar, S. 1942, ApJ, 96, 161

The Golden Gift of Red Giants

Arlette Noels-Grotsch and Andrea Miglio

Chapter 5

Red Giant Phase

5.1 Ages of Red Giants

Why are Red Giants a golden gift to asteroseismologists?

Equation (4.9) is really the answer to that question if we add the fact that the rapid crossing of the Hertzsprung gap only slightly increases the ages whatever the amount of extra-mixing above the convective core during MS.[1] We can write

$$\tau_{RG} \sim \tau_{SC}. \tag{5.1}$$

This is true for low luminosity RGs but also for the whole Red Giant Branch (RGB) since the time spent on the RGB up to the tip is much smaller than the MS lifetime.[2] Figure 5.1 shows the age–mass–metallicity relation for MS stars (left panel) and RG stars (right panel) in a synthetic population (computed with TRILEGAL (Girardi et al. 2005) representative of thin-disk stars observed by CoRoT (Baglin et al. 2006). It illustrates the drastic difference between MS and RG stars. For an MS star of given mass and metallicity, the age dispersion is enormous while for a RG star it is extremely small. The ages displayed in the right panel are indeed a good proxy for the ages at the Schönberg–Chandrasekhar mass limit, τ_{SC}, in the Post Main Sequence phase.

---

[1] This is true for a moderate overshooting with $\alpha \lesssim \alpha_{ov,SC\text{-}TAMS}$ (see Section 4.2).

[2] In low-mass stars the amount of hydrogen burnt in the H-shell on the RGB is of the same order than during MS but the luminosity is much higher, which drastically decreases the time span of this phase.

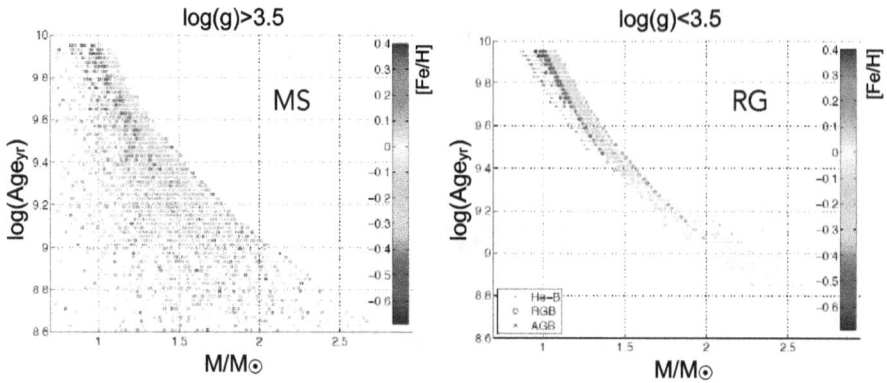

Figure 5.1. Age–mass–metallicity relation for main-sequence stars (left panel) and red giants (right panel) in a synthetic population (Girardi et al. 2005) representative of thin-disk stars observed by CoRoT in the LRc01 field (Baglin et al. 2006). The evolutionary state of giants is marked with a different symbol: dots (stars in the core helium burning phase), crosses (Asymptotic Giant Branch stars), and open circles (stars on the RGB). Reproduced from Miglio (2012), with permission from Springer Nature.

5.2 Ascending the RGB

Why is L increasing on the RGB?

After crossing the Hertzsprung gap, our star reaches a track close to the Hayashi track (see Section 2.1) and from now on the evolution proceeds along this track with an increasing luminosity and a larger and larger radius. The outer layers are thus expanding and cooling, which allows a deepening of the convective envelope.

As a result of the drastic contraction of the helium core, the degeneracy level increases and central layers can once more evolve toward an isothermal structure.[3] This is done through a transient cooling of the most inner layers as can be seen in Figure 4.1 (see also Section 4.1).

The stellar structure is now as follows:

- a dense quasi-isothermal helium core,
- a H-shell whose outward motion increases the mass of the helium core,
- a thinner and thinner intermediate radiative zone, and
- a deepening hydrogen-rich convective envelope whose lower boundary comes closer and closer to the H-shell.

Figure 5.2 schematically shows the H-shell located immediately above radius $R_{\rm He}$ (radius of the helium core) and underneath r_0. From the condition of hydrostatic equilibrium, we can write for the pressure at radius r inside the H-shell

---

[3] As seen in Section 4.2 there is no Schönberg–Chandrasekhar mass limit in the presence of degeneracy.

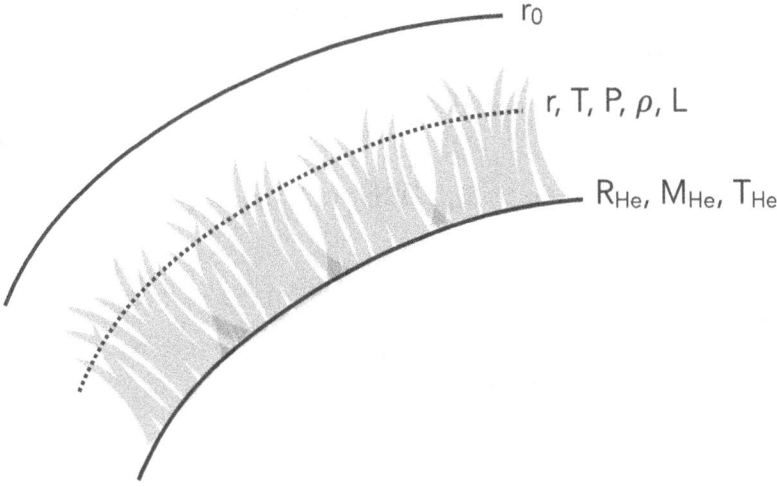

Figure 5.2. Schematic representation of the H-shell located above the helium core and underneath r_0. The mass, radius, and temperature at the border of the core are, respectively, M_{He}, R_{He}, and T_{He}.

$$P = P(r_0) + \int_r^{r_0} \frac{G\rho m}{r^2}dr \sim \frac{GM_{He}}{R_{He}} \int_{R_{He}/r_0}^{R_{He}/r} \rho \, d\left(\frac{R_{He}}{r}\right) \sim \frac{GM_{He}}{R_{He}}\rho \qquad (5.2)$$

since $P(r_0)$ is much smaller than P owing to the rapid outward drop in pressure between r and r_0.

All the energy is produced inside the H-shell, which translates into

$$L = 4\pi \int_{R_{He}}^r \varepsilon\rho r^2 dr \sim X_{CNO} \, \rho^2 T_{He}^\nu R_{He}^3. \qquad (5.3)$$

As matter is non-degenerate inside the H-shell this becomes

$$L \sim X_{CNO} \, \rho^2 \left(\frac{M_{He} \, \mu}{R_{He}}\right)^\nu R_{He}^3 \sim X_{CNO} \, \rho^2 (M_{He} \, \mu)^\nu R_{He}^{3-\nu} \qquad (5.4)$$

where μ is the mean molecular weight within the H-shell.

From Equation (2.2), we can write

$$L \sim \frac{R_{He} T_{He}^4}{\rho} \sim \frac{(M_{He}\mu)^4 R_{He}^{-3}}{\rho} \qquad (5.5)$$

and by equating Equations (5.4) and (5.5) we obtain the following expression for the density at level r in the H-shell:

$$\rho \sim (M_{He}\mu)^{\frac{4-\nu}{3}} R_{He}^{\frac{\nu-6}{3}} X_{CNO}^{1/3}, \qquad (5.6)$$

which gives after reframing Equation (5.5)

$$L \sim (M_{He} \, \mu)^{\frac{\nu-4}{3}+4} R_{He}^{\frac{6-\nu}{3}-3} X_{CNO}^{1/3} \sim (M_{He} \, \mu)^7 R_{He}^{-\frac{16}{3}} X_{CNO}^{1/3} \quad \text{if } \nu = 13. \qquad (5.7)$$

Now, the dense helium core is degenerate, and from the dimensional expression of Equation (2.14), we have

$$R_{He} \sim M_{He}^{-1/3} \tag{5.8}$$

which leads to

$$L \sim (M_{He}\,\mu)^7 M_{He}^{16/9} X_{CNO}^{1/3} \sim M_{He}^9 \mu^7 X_{CNO}^{1/3} \tag{5.9}$$

$$T_{He} \sim \frac{\mu M_{He}}{R_{He}} \sim M_{He}^{\frac{4}{3}} \mu \tag{5.10}$$

$$\frac{dM_{He}}{dt} \sim \frac{L}{X Q_{CNO}} \sim M_{He}^9 \mu^7 X_{CNO}^{1/3} / X \tag{5.11}$$

$$\frac{dT_{He}}{dt} \sim M_{He}^{9+1/3} \mu^8 X_{CNO}^{1/3} / X \tag{5.12}$$

where Q_{CNO} is the energy produced by the CNO cycle when transforming four protons into a helium nucleus.

This clearly shows that as hydrogen is burnt into the H-shell, leading to a progressively larger and larger helium core, the luminosity increases. Since the effective temperature is quasi-constant on the RGB[4] this means that the radius must increase and the outer layers cool down. As a result, the convective envelope goes on expanding at the beginning of the ascension of the RGB.

It is also interesting to notice that Equation (5.10) is similar to Equation (3.13) and shows that the mean temperature in the degenerate quasi-isothermal helium core increases with the growth of the core. However, we have seen in Section 2.1 that a contracting degenerate gas should cool down. This apparently contradictory result comes from the fact that M_{He} is not kept constant but rather increases along the RGB.

This value of the He core temperature is also that of the H-shell and it must increase with time as the He core grows. To produce a larger luminosity with a higher temperature in the H-shell implies a thinner H-shell. *Following the growth of the helium core, the H-shell becomes hotter and thinner.* This heating of the H-shell is in agreement with Equation (3.13) since the mean molecular weight increases in hydrogen burning layers located at the edge of a larger and larger helium core mass with a smaller and smaller radius.

5.3 First Dredge-Up and RGB Bump

Why is there a first dredge-up?

After the vanishing of the convective core at the TAMS, hydrogen burns in the H-shell, which progressively moves outward as the helium core mass grows. This is

---

[4] The Hayashi track (see Section 2.1) is now traveled in the opposite way than during PMS.

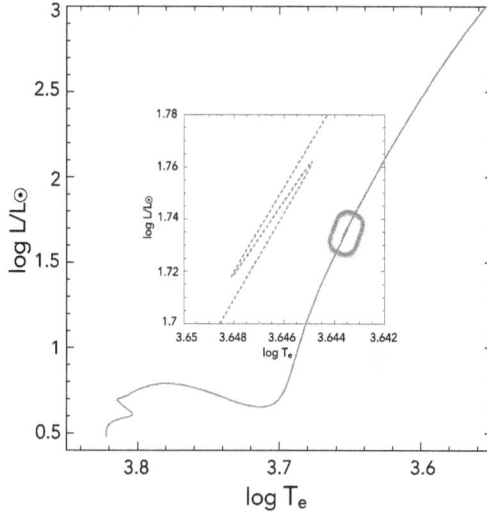

Figure 5.3. HR diagram of a 1.3 $M_\odot$ star during its RGB ascension. The inset shows a zoom on the RGB bump.

accompanied by an expansion of the outer layers and a deepening of the cooler and cooler convective envelope (see Section 5.2). The base of the envelope eventually penetrates layers that have been sufficiently affected by nuclear burning to have a chemical composition different from the surface values (see Section 3.2, Figure 3.2). From now on the chemical composition of the surface layers starts to change as a result of the *first dredge-up*, which brings to the surface matter processed by hydrogen burning with a decrease of the C^{12}/N^{14} and C^{12}/C^{13} ratios, and a lowering of Li^7 and Be^8.

During its ascension of the RGB our star encounters a zigzag, which is known as the RGB bump (Figure 5.3) first discovered in evolutionary computations of the RGB by Sweigart & Gross (1978). This bump is observed as an accumulation of stars in HR or color–magnitude diagram in GCs (Lagioia et al. 2018) as well as in dSph galaxies (see Figure 5.4; Monelli et al. 2010).

What is the physical origin of the RGB bump?

The physical reason for the presence of this bump can be found in Figure 5.5 (Thomas 1967). The left panel shows an À la Kippenhahn diagram for a 1.3 $M_\odot$ Pop II star where line shaded zones are nuclear burning regions and curved symbols mark convective layers. The right panel is the HR diagram for the star depicted in the left panel. The deepening of the convective envelope, responsible for the first dredge-up, is clearly seen. At the same time, the thinner and thinner H-shell moves outwards and comes closer and closer to the convective boundary of the envelope while the helium core mass grows accordingly. This implies an increase of

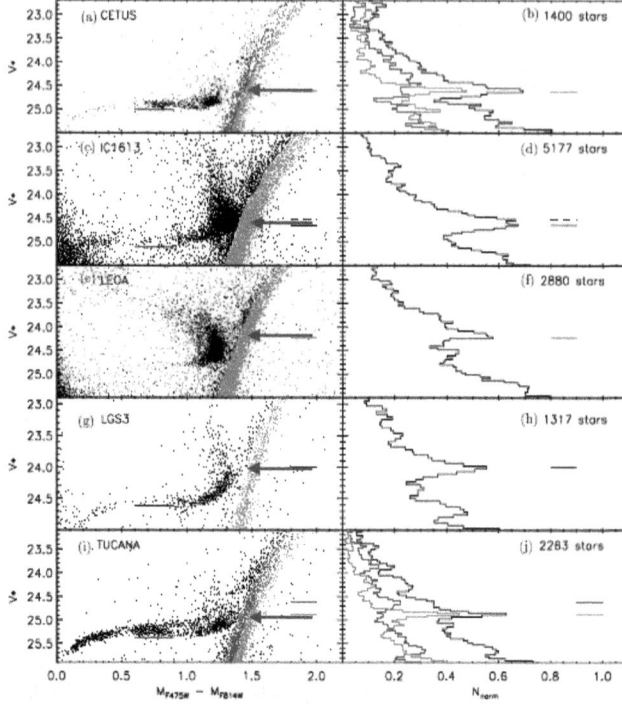

Figure 5.4. Left panels: the observed CMDs for five dSph galaxies. The arrows mark the RGB bump. Right panels: the observed differential RGB luminosity functions. Reproduced from Monelli et al. (2010). © 2010. The American Astronomical Society. All rights reserved.

temperature in the helium core as well as in the H-shell and the layers close by (see Equation (5.10)).

The heating of these layers as a result of the growth of the He core allows hydrogen combustion in the H-shell but also keeps the opacity low enough to prevent convection and the convective envelope eventually recedes. However, while receding, the convective envelope leaves a discontinuity in the chemical composition. Once the H-shell comes into contact with this discontinuity, it suddenly operates in a matter much richer in hydrogen. Since matter is non-degenerate in the H-shell, the thermostatic pressure control mechanism (see Section 3.2.1) acts and the H-shell expands and cools down. This can be easily seen by writing the following relation similar to Equation (3.13) (see also Equation (5.10))

$$T_{\text{shell}} \sim \frac{m_{\text{shell}}}{r_{\text{shell}}} \mu_{\text{shell}} \sim \frac{M_{\text{He}}}{R_{\text{He}}} \mu_{\text{shell}} \qquad (5.13)$$

where the subscript shell relates to typical conditions within the H-shell. Since the helium core is contracting (Equation (5.8)) while growing in mass, T_{shell} increases as long as μ_{shell} does not change too drastically. Once the X discontinuity is

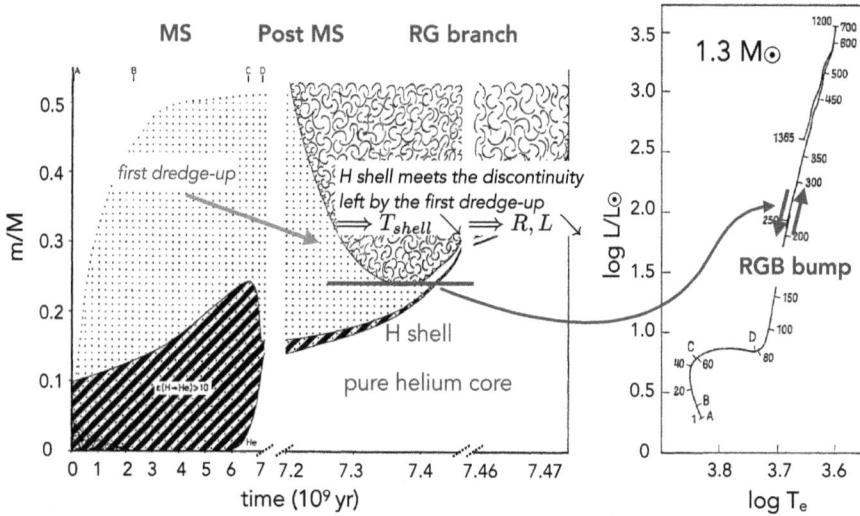

Figure 5.5. Left panel: À la Kippenhahn diagram showing the structural evolution of a 1.3 $M_\odot$ Pop II star. Line shaded zones are nuclear burning regions and curved symbols mark convective layers. Right panel: HR diagram for the 1.3 $M_\odot$ Pop II star illustrated in the left panel. Reproduced from Thomas (1967), with permission from Springer Nature.

encountered, this is not the case anymore since μ_{shell} suddenly decreases and the H-shell cools off. This is immediately followed by a global contraction and a decrease in luminosity before resuming the ascension of the RGB (see also Christensen-Dalsgaard 2015).

The theoretical RGB bump is more luminous, by 0.2–0.4 mag, than observed in GCs. To decrease the RGB bump down to the observed value requires deviations from standard models. An overshooting of $\alpha_{\mathrm{ov}} = 0.5 H_P$ below the convective envelope (undershooting), which allows the H discontinuity to be located at a smaller mass fraction in the star, allows the encounter of the H-shell with the H discontinuity to occur earlier, i.e., at a lower luminosity (Alongi et al. 1991; see also Khan et al. 2018 for an analysis including seismic data). Such a decrease in luminosity can also be achieved by increasing the opacity below the convective boundary, either with diffusion or by increasing the abundances of α-elements. Taking into account α-enhanced metallicities as well as undershooting markedly improves the location of the RGB bump in the GC 47 Tuc (Fu et al. 2018).

What is the effect of the mass?

Figure 5.6 (left panel) shows the location and amplitude of the RGB bump for different values of the stellar mass. As the mass increases the bump becomes less and less important. However, for masses above 2 $M_\odot$, the reverse is observed. The RGB bump conditions for these masses are in fact already typical of core helium burning.

The evolution of the temperature gradients ∇_{rad}, ∇_{ad}, ∇_T is shown in Figure 5.7. The hydrogen abundance, X, and the helium abundance, Y, are the green and red

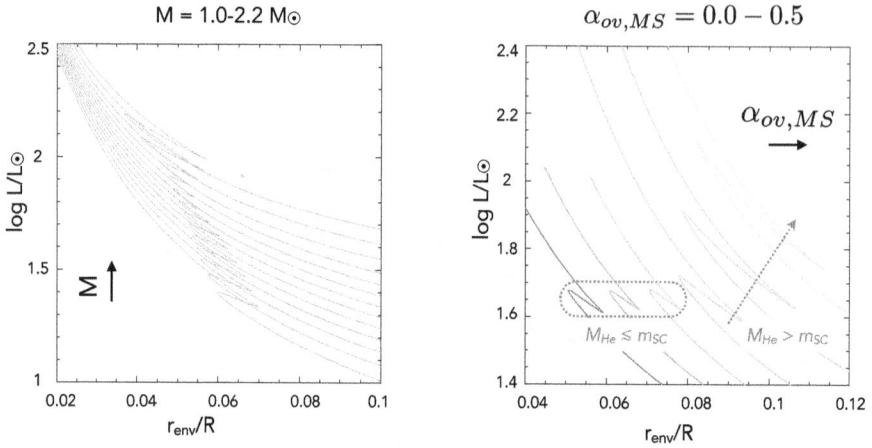

Figure 5.6. Luminosity vs. fractional radius at the base of the convective envelope, r_{env}/R, near and during the RGB bump for different values of the stellar mass (left panel) and for different values of the overshooting parameter, $\alpha_{ov,MS}$, adopted to compute the main-sequence phase of a 1.3 $M_\odot$ (right panel).

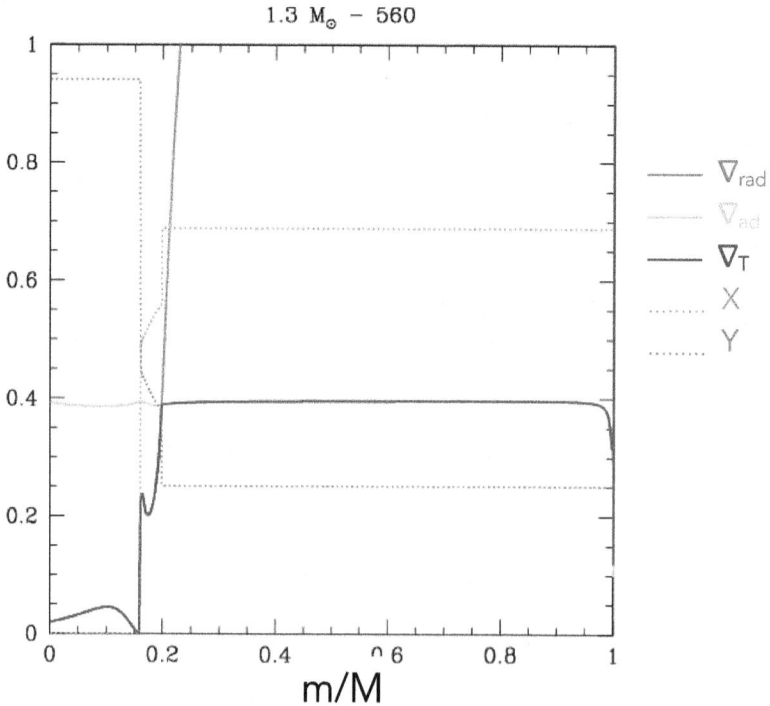

Figure 5.7. Figure showing the penetration of the convective envelope in the Post MS phase and the physical conditions responsible for the bump. The temperature gradients ∇_{rad}, ∇_{ad}, ∇_T are shown in red, cyan, and blue, respectively. The hydrogen abundance, X, and the helium abundance, Y, are displayed as green and red dotted curves, respectively. This figure corresponds to Movie5 in the supplementary Keynote and pptx files, accessible at http://doi.org/10.1088/2514-3433/adcf15.

dotted curves, respectively. The penetration of the convective envelope, which is responsible for the first dredge-up, can easily be followed. The near encounter of the convective envelope lower limit and the H-shell is also clearly visible in the figure.

What is the effect of MS extra-mixing?

The influence of convective core overshooting during MS on the location of the RGB bump in the HR diagram is interesting since it is related to the Schönberg–Chandrasekhar mass limit (see Section 4.2). For values of the overshooting parameter such that the He core mass at the TAMS is smaller than or equal to this mass limit, the mass of the He core at the bottom of the RGB is very similar (see Section 4.2). It immediately ensues that the properties of the RGB must be about the same, in particular the amplitude and the mean luminosity of the bump are very similar. The situation changes drastically when M_{He} at TAMS is already larger than m_{SC}. The helium core mimics that of a more massive star and the mean luminosity of the bump increases accordingly as can be seen in Figure 5.7 (right panel) where the luminosity versus the fractional radius of the convective envelope is drawn for different values of the overshooting parameter $\alpha_{ov,MS}$ color coded in the legend.

5.4 List of Questions

Why are Red Giants a golden gift to asteroseismologists?
Why is L increasing on the RGB?
Why is there a first dredge-up?
What is the physical origin of the RGB bump?
What is the effect of the mass?
What is the effect of MS extra-mixing?

References

Alongi, M., Bertelli, G., Bressan, A., & Chiosi, C. 1991, A&A, 244, 95

Baglin, A., Auvergne, M., Barge, P., et al. 2006, Proc. of "The CoRoT Mission Pre-Launch Status – Stellar Seismology and Planet Finding" (ESA SP-1306) (eds) M. Fridlund, A. Baglin, J. Lochard and L. Conroy (Paris, ESA) 33

Christensen-Dalsgaard, J. 2015, MNRAS, 453, 666

Fu, X., Bressan, A., Marigo, P., et al. 2018, MNRAS, 476, 496

Girardi, L., Groenewegen, M. A. T., Hatziminaoglou, E., & da Costa, L. 2005, A&A, 436, 895

Khan, S., Hall, O. J., Miglio, A., et al. 2018, ApJ, 859, 156

Lagioia, E. P., Milone, A. P., Marino, A. F., et al. 2018, MNRAS, 475, 4088

Miglio, A. 2012, ApSSP, 26, 11

Monelli, M., Cassisi, S., Bernard, E. J., et al. 2010, ApJ, 718, 707

Sweigart, A. V., & Gross, P. G. 1978, ApJS, 36, 405

Thomas, H.-C. 1967, ZA, 67, 420

Arlette Noels-Grotsch and Andrea Miglio

Chapter 6

Core Helium Burning

6.1 Onset of Core Helium Burning

6.1.1 Low Mass Stars—$M \lesssim 1.8 \; M_\odot$[1]

Why is the temperature increasing in a contracting degenerate core?

After the formation of a quasi-isothermal helium core, the mean core temperature resumes its increasing behavior according to Equation (5.10) as can be seen in Figure 4.1. The dotted curve in Figure 4.1 schematically shows the cooling that should be expected if the mass of the helium core were kept constant. This is a very important point, which is not in contradiction with the Viriel theorem (Section 2.2): *The temperature increase of the helium core in a low mass star ascending the RGB is not due to the energy released by contraction but is the result of the growth of the helium core mass.*

What is the mass of the helium core at the onset of helium burning in low mass stars?

Figure 6.1 shows the evolution of the central conditions in the minimum mass for a non-degenerate pure helium star (0.33 $M_\odot$, green dotted curve) to reach helium burning temperatures ($\sim 10^8$ K). The dotted blue curve shows that a lower star mass cannot start burning helium while the full green and blue curves show that with a growing helium core mass, the central temperature resumes its increasing behavior. When a value of about 0.48 $M_\odot$ is reached, conditions are fulfilled to start core helium burning.

This value of $\sim 0.48 \; M_\odot$ is the mass of the core that *all low mass stars* must form to start burning helium. This onset of nuclear reactions in a degenerate core gives rise to the so-called *Helium Flash*.

---

[1] See Foreword—Note 4.

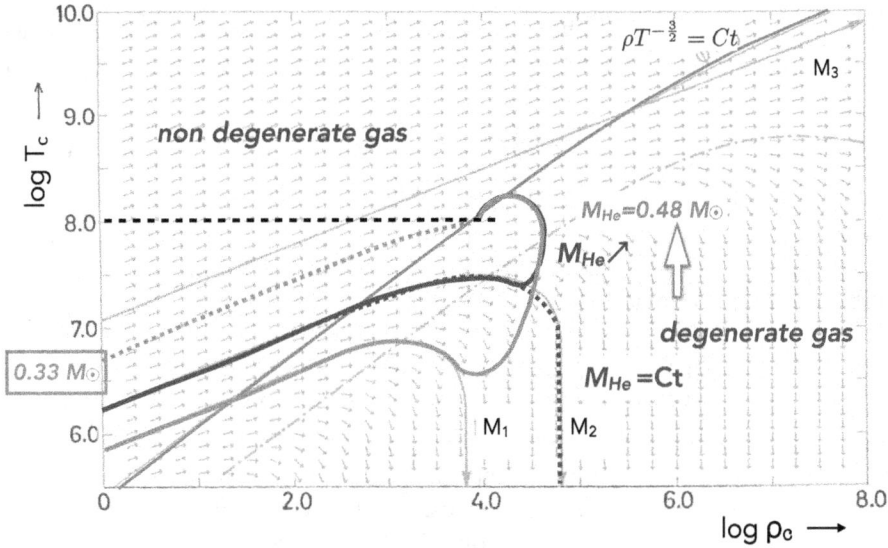

Figure 6.1. log T_c vs log ρ_c diagram. The green dotted line shows the evolution of the central conditions in the non-degenerate minimum mass for He burning. The full blue and green curves relate to degenerate He core growing in mass. Reproduced from Kippenhahn & Weigert (1994), with permission from Springer Nature.

The evolution leading to the He flash in a low mass star ($M \lesssim 1.8\ M_\odot$) is as follows[2]:

- Contraction of the quasi-isothermal degenerate He core of increasing mass and increasing temperature.
- Nuclear burning in the surrounding non-degenerate H-shell.
- Expanding outer layers and increasing luminosity.
- At the threshold helium core mass value of $\sim 0.48\ M_\odot$, onset of He burning in the degenerate core.

Why is there a flash at the onset of helium burning?

Helium burning transforms helium into mostly carbon and oxygen through the following reactions:

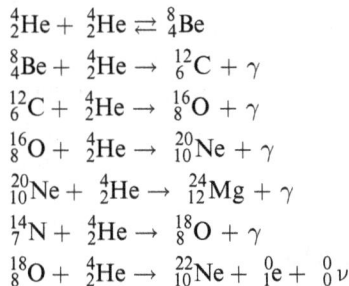

$$^{4}_{2}\text{He} + ^{4}_{2}\text{He} \rightleftarrows ^{8}_{4}\text{Be}$$

$$^{8}_{4}\text{Be} + ^{4}_{2}\text{He} \rightarrow ^{12}_{6}\text{C} + \gamma$$

$$^{12}_{6}\text{C} + ^{4}_{2}\text{He} \rightarrow ^{16}_{8}\text{O} + \gamma$$

$$^{16}_{8}\text{O} + ^{4}_{2}\text{He} \rightarrow ^{20}_{10}\text{Ne} + \gamma$$

$$^{20}_{10}\text{Ne} + ^{4}_{2}\text{He} \rightarrow ^{24}_{12}\text{Mg} + \gamma$$

$$^{14}_{7}\text{N} + ^{4}_{2}\text{He} \rightarrow ^{18}_{8}\text{O} + \gamma$$

$$^{18}_{8}\text{O} + ^{4}_{2}\text{He} \rightarrow ^{22}_{10}\text{Ne} + ^{0}_{1}e + ^{0}_{0}\nu$$

---

[2] For a more detailed description of the He flash, see, for example, Salaris & Cassisi (2006).

We have seen in Section 3.2.1 that adding heat to a degenerate gas induces a heating ($c^* = c_v > 0$). Such a heating increases the energy production rate, which is extremely temperature sensitive ($\nu_{3\alpha} \sim 40$) and can be written

$$\varepsilon_{3\alpha} \sim Y^3 \rho^2 T^{40} \text{ at } T \sim 10^8 \text{ K.} \tag{6.1}$$

This increase in the energy production in turn brings more heat, and a thermal runaway, i.e., a flash, ensues. This can also be understood with the following argument similar to what we did in Section 3.2.1. In order to maintain hydrostatic equilibrium we must have

$$P_c = K \left(\frac{\rho_c}{\mu_e} \right)^{5/3} (\text{Equation} \quad (2.14)) = \int_0^M \frac{Gm}{4\pi r^4} \, dm. \tag{6.2}$$

An increase in temperature, which would be prevented by an expansion in a non-degenerate gas, has no impact on the central pressure, and heating can occur without any expansion response from the core layers. In other words, there is no thermostatic pressure control mechanism (see Section 3.2.1) in a degenerate gas and a thermal runaway—*a flash*—ensues. These physical conditions are once more typical of a secular instability, i.e., a perturbation of thermal equilibrium without any effect on hydrostatic equilibrium (see Section 4.3).

What are the core temperature and density behaviors during the flash?

In low mass stars the problem is slightly tricky since the maximum temperature is not found at the center. With the increasing degeneracy level during the RGB ascension, electron conduction mostly operates the energy transfer, which tends to decrease the temperature gradient. On the other hand, the production of neutrinos (Schinder et al. 1987) provides a cooling the most important as the density is larger, i.e., in the layers closest to the center. The net impact is the formation of an inverted temperature profile with a maximum temperature (T_{max}) reached at a mass of about 0.2 $M_\odot$ within the helium core (see Montalbán & Noels 2013 and references therein). The He flash starts off-center with a huge production of energy ($L \sim 10^{10} L_\odot$) in a very short time interval.

However, as the temperature severely rises, the degeneracy level decreases and a thermostatic pressure control mechanism partially acts (see Section 3.2.1). This triggers an expansion and a cooling down of the burning layers, which in turn leads to an increase of the degeneracy level. This is a kind of recurrent process with a decrease/increase of degeneracy accompanied by a heating/cooling of layers closer and closer to the center through a series of secondary flashes. The maximum temperature is also found closer and closer to the center until it eventually reaches the center. This can be seen in Figure 6.2 where the evolution of low mass stars ($M \lesssim 1.8 \ M_\odot$) in the log T_{max} versus log ρ_c plane is illustrated by the red dotted curve (Montalbán & Noels 2013). The recurrent process is marked by successive loops of decreasing amplitude since each loop has a mean degeneracy level smaller than the preceding loop. Degeneracy is eventually lifted at the center and the "oscillating" stellar core converges toward a non-degenerate helium burning core of about

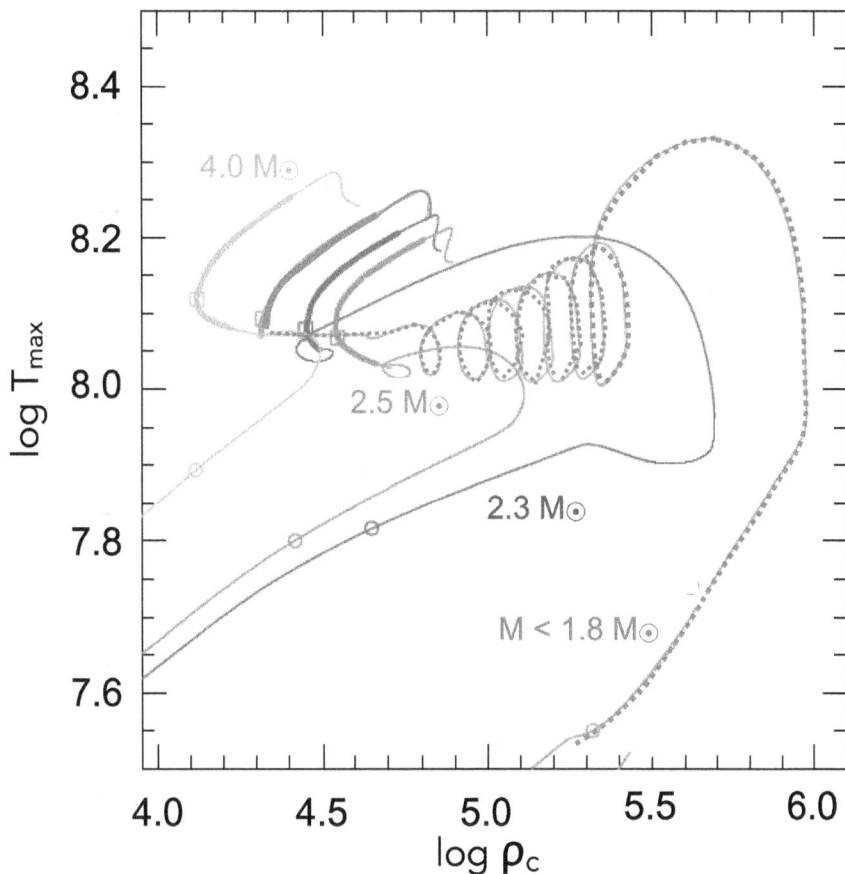

Figure 6.2. $\log T_{\max}$ versus $\log \rho_c$ diagram for core helium burning stars in models computed with ATON Ventura et al. (2008). For stars more massive than $\sim 1.8 M_\odot$ $T_{\max}$ is the central temperature ($T_{\max} = T_c$). Low mass stars ($M \lesssim 1.8\ M_\odot$) follow a unique red dotted curve. Adapted from Montalbán & Noels (2013). CC BY 2.0.

$0.48\ M_\odot$. Interestingly enough, and for the reasons stated above, all stars of masses $\lesssim$ $1.8\ M_\odot$ follow the same path in the $\log T_{\max}$ versus $\log \rho_c$ plane.

Such loops are also visible in the HR diagram (Figure 6.3) since each time the thermostatic pressure control mechanism acts, it induces a global response of the envelope as we have seen in Section 3.2.1.

The computation of models undergoing the helium flash is very delicate and time consuming. That is the reason why most stellar evolution codes have developed strategies to skip that phase and start the core helium burning phase with post flash models. Specific assumptions must be made, among which the most important are:

- the mass fraction of the non-degenerate helium core, which is indeed very close to the mass of the degenerate helium core at the onset of the flash,
- the amount of carbon produced during the flash, which can be estimated by matching the gravitational potential energy of the helium core and the total nuclear energy production through the 3α reactions during the flash.

A relaxation procedure is then used to lift the degeneracy of the core.

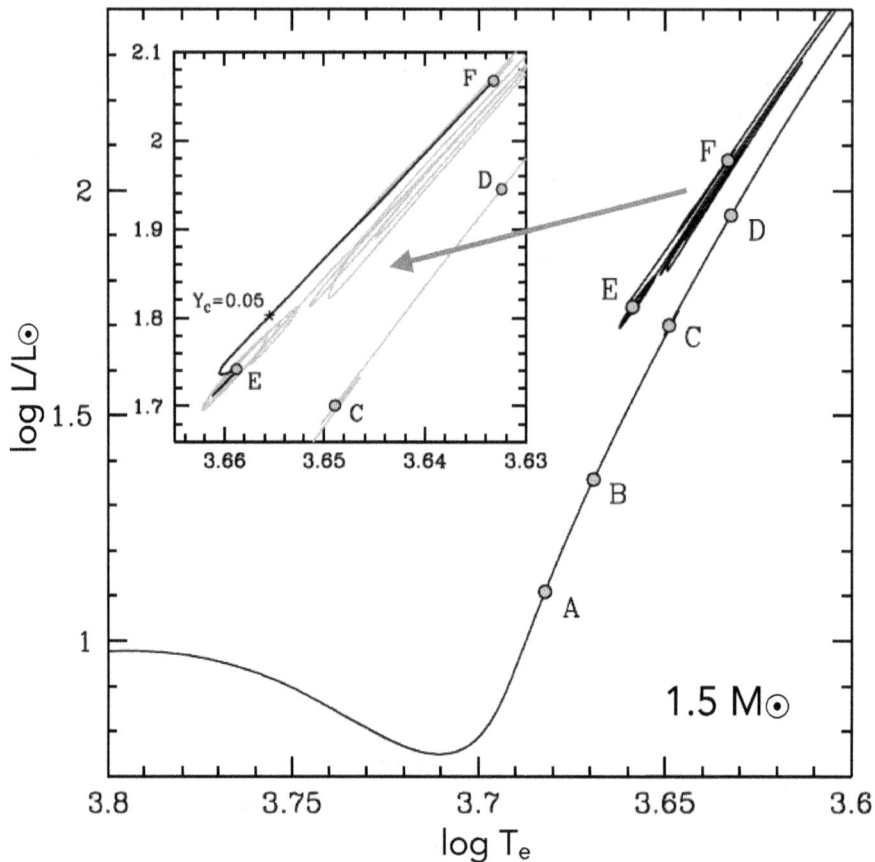

Figure 6.3. Evolutionary track of a 1.5 $M_\odot$ star during the red giant phase, helium flash, and core helium burning. Adapted from Montalbán & Noels (2013). CC BY 2.0.

6.1.2 Intermediate Mass Stars—$M \gtrsim 2.4\ M_\odot$[3]

What is the behavior of an intermediate mass star?

Stars more massive than ∼2.4 $M_\odot$ follow an entirely different channel during their red giant phase. Their helium core is non-degenerate and heats up as a result of its contraction (Section 2.2) while ascending the RGB. The threshold temperature for helium burning is eventually achieved without any important raise in the mass of the helium core, which can still remain close to the Schönberg–Chandrasekhar mass limit, i.e., 0.1 M (Section 4.2) with however a minimum mass of the helium core of

---

[3] See Foreword—Note 4.

the order of $M_{He} \sim 0.33\ M_\odot$ (Section 3.1). This minimum core mass is encountered in a star of 2.4 $M_\odot$, which is the lower mass limit of *intermediate mass stars*.

The evolution leading to He burning in an intermediate mass star ($M \gtrsim 2.4\ M_\odot$) is as follows:

- Contraction of the non-degenerate helium core of mass $M_{He} \sim 0.1\ M$.
- Burning in the non-degenerate H-shell, which slightly increases the mass of the He core.
- At the threshold temperature value of $\sim 10^8$ K, onset of core He burning in a helium core of the order of 0.1 M, but with $M_{He} \gtrsim 0.33\ M_\odot$.

The evolution in the $\log T_c$ versus $\log \rho_c$ diagram is illustrated in Figure 6.2 where a specific curve is allotted to each stellar mass. The bends that can be seen in these curves are the signatures of the thermostatic pressure control mechanism operating in a non-degenerate gas (Section 3.2.1).

6.1.3 Transition Mass

What is the mass of the helium core at the onset of core helium burning?

Figure 6.4 shows the helium core mass as a function of the total stellar mass for models computed without overshooting (black curve). Stars of masses below 1.8 $M_\odot$ start helium burning with a flash in a helium core mass of about 0.48 $M_\odot$. As we go from stars with $M \sim 1.8\ M_\odot$ to higher masses on the RGB, we progressively find H-shell burning stars with a lower degree of electron degeneracy in their cores and helium is eventually quietly ignited in the core. It ensues that the mass of the helium core at the onset of helium burning progressively decreases, reaching a minimum at a stellar mass which slightly depends on the initial chemical composition and on the amount of extra mixing during MS. After this minimum, the He core grows again with mass, following the increasing size of the core at the Schönberg–Chandrasekhar mass limit ($\sim 0.1\ M$, see Section 4.2).

This minimum helium core mass is close to 0.33 $M_\odot$, which is also the minimum mass of a pure helium star that can start burning helium in non-degenerate conditions (see Section 3.1). The transition mass, m_{trans}, i.e., the total stellar mass housing this minimum helium core mass, is close to 2.4 $M_\odot$. Its definition can be slightly rephrased as: *The transition mass ($\sim 2.4\ M_\odot$) is the mass of the star that can host the minimum mass of a pure non-degenerate helium burning star ($\sim 0.33\ M_\odot$).* It fixes the limit between low and intermediate mass stars on a purely physical basis. Like all the mass limits discussed so far, the transition mass is a very robust notion based on solid physical concepts.

Figure 6.5 shows the mass of the helium core at the onset of core helium burning computed with CLES and AGSS09 abundances (left panel). The masses have been obtained from evolutionary models at the onset of the helium flash and a relaxation procedure to lift the degeneracy (see Section 6.1.1).

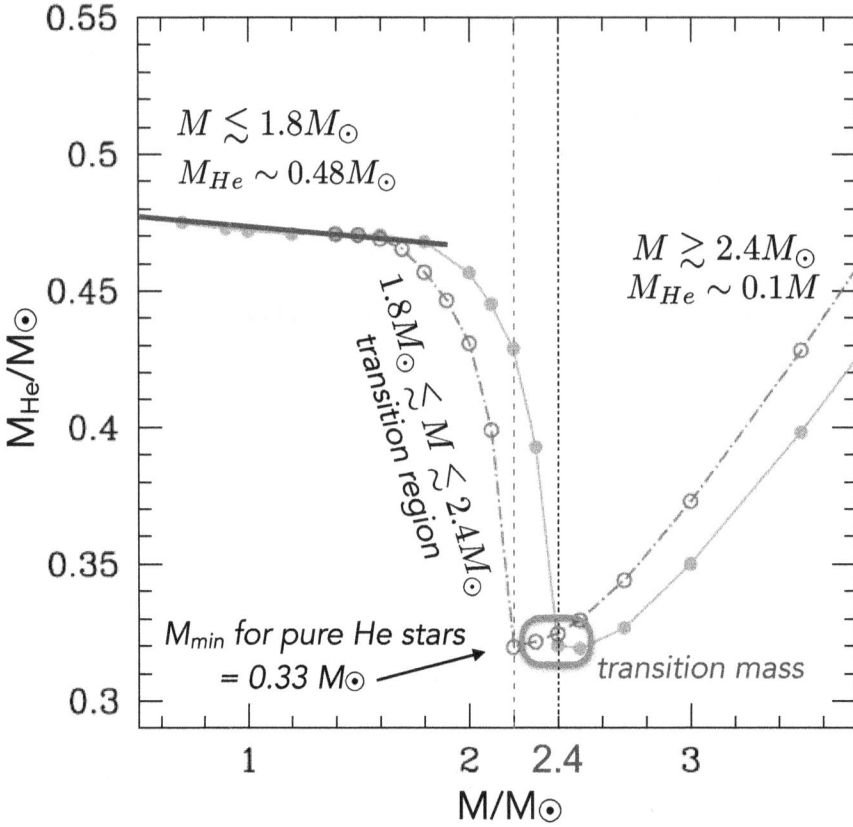

Figure 6.4. Mass of the helium core at the onset of helium burning as a function of the total stellar mass. The black/red dotted curve has been computed without/with MS diffusive overshooting ($f_{ov,MS} \sim 0.02$) (Ventura et al. 2008) and shows the influence of MS extra mixing on the transition mass. Adapted from Montalbán & Noels (2013). CC BY 2.0.

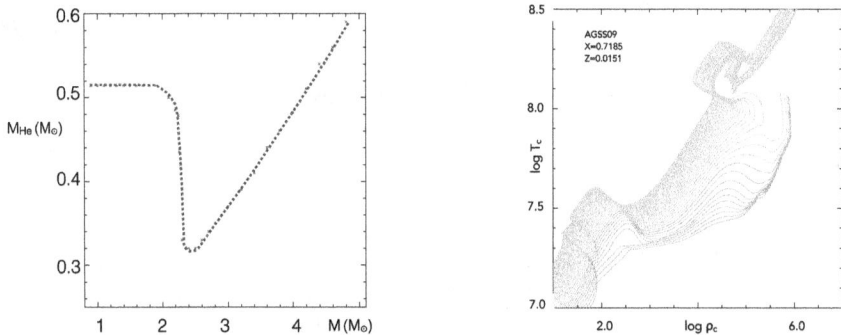

Figure 6.5. Mass of the helium core at the onset of core helium burning (left panel) obtained from CLES models computed with AGSS09 abundances at the onset of helium flash (right panel)

What can affect the transition mass?

6.1.3.1 Extra-Mixing

What is the impact of MS extra mixing?

Figure 6.4 also shows the behavior of the helium core mass with respect to the total mass for models computed with MS diffusive overshooting ($f_{ov,MS} = 0.02$)[4] Ventura et al. (2008). Adding an extra mixing during MS slightly increases the MS luminosity and the RG evolution mimics that of a "slightly more massive" star, which lowers the transition mass accordingly (2.2 $M_\odot$ instead of 2.4 $M_\odot$). Models computed with instantaneous $\alpha_{ov,MS} = 0.0$–0.1–0.2 are shown in Figure 6.6 for a range of masses covering 1.0–4.9 $M_\odot$ and the chemical composition given in the legend. Although the curves look very similar they are related to different stellar masses and m_{trans} changes accordingly, the larger $\alpha_{ov,MS}$, the smaller m_{trans}.

On the other hand, *for moderate values of the overshooting parameter*, the mass of the helium core is about the same at the RGB bottom, whether or not an extra mixing is adopted (see Section 4.2). From Equation (5.12), the time interval to reach the threshold temperature for helium burning is not affected and M_{He} at the onset of helium burning for stars undergoing a flash remains unchanged (see Figure 6.4). However, if the amount of extra mixing is large enough for a mass of the helium core at TAMS larger than the Shönberg–Chandrasekhar mass limit, this time interval decreases and M_{He} at the onset of helium burning decreases accordingly.

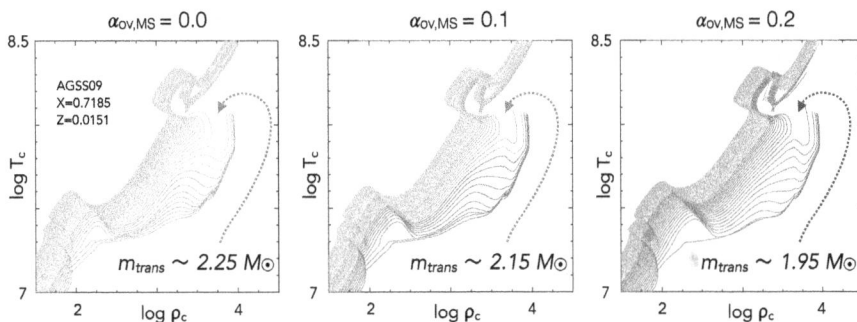

Figure 6.6. log T_c *versus* log ρ_c diagram for a range of masses covering 1.0–4.9 $M_\odot$ computed with the chemical composition given in the legend and different values of the overshooting parameter (0.0, 0.1, 0.2). The arrows show the separation between stars undergoing a helium flash and those starting core helium burning in non-degenerate conditions, which happens at a value m_{trans} all the smaller than $\alpha_{ov,MS}$ is larger.

---

[4] A similar impact on the MS phase with an instantaneous mixing would require $\alpha_{ov,MS} \sim 0.2$ (see Equation (3.18)).

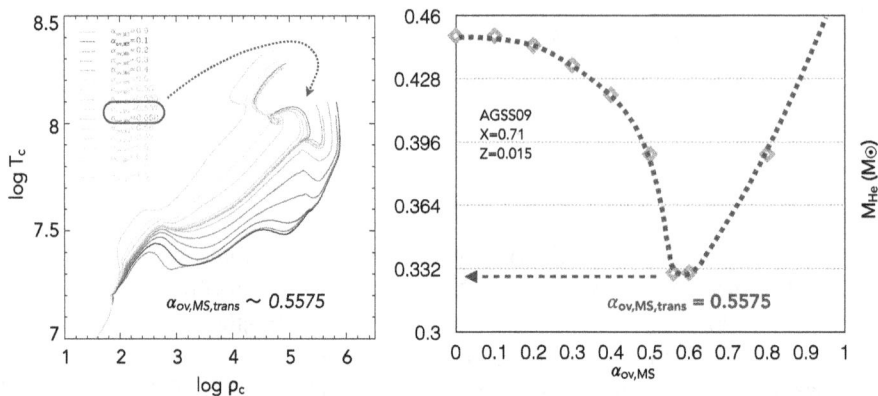

Figure 6.7. Left panel: $\log T_c$ vs $\log \rho_c$ diagram for a star of 1.3 $M_\odot$ computed with the chemical composition given in the legend and different values of the instantaneous MS overshooting parameter (from 0.0 to 1.5). The arrow shows the separation between stars undergoing a helium flash and those starting core helium burning in non-degenerate conditions, which happens at $\alpha_{\rm ov,MS,trans} \sim 0.5575$. Right panel: Mass of the helium core at the onset of helium burning for a star of 1.3 $M_\odot$ as a function of $\alpha_{\rm ov,MS}$. For a value of $\alpha_{\rm ov,MS,trans} \sim 0.5575$, the mass of the helium core is of the order of 0.33 $M_\odot$.

An Exercise

As an exercise, we can fix the stellar mass at 1.3 $M_\odot$ and vary the instantaneous overshooting parameter $\alpha_{\rm ov,MS}$ from 0.0 to 1.5. The left panel of Figure 6.7 is a $\log T_c$ versus $\log \rho_c$ diagram for a star of 1.3 $M_\odot$ computed with a large range of instantaneous $\alpha_{\rm ov,MS}$ listed in the legend. The larger $\alpha_{\rm ov,MS}$, the more massive star our 1.3 $M_\odot$ mimics. Above a transition value $\alpha_{\rm ov,MS,trans} \sim 0.5575$, our 1.3 $M_\odot$ starts core helium burning in non-degenerate conditions. The right panel of Figure 6.7 shows the mass of the helium core at the onset of core helium burning for this range of $\alpha_{\rm ov,MS}$. At the transition $\alpha_{\rm ov,MS,trans} \sim 0.5575$, the mass of the helium core is close to 0.33 $M_\odot$. Note the lack of an extended plateau due to the fact that for $\alpha_{\rm ov,MS}$ larger than about 0.2, $M_{\rm He}$ at TAMS is already larger than $m_{\rm SC}$, which means that no isothermal core can form and $M_{\rm He}$ at the bottom of RGB is accordingly larger than for smaller values of $\alpha_{\rm ov,MS}$. The mass increment required to reach the threshold temperature for helium burning is thus smaller and $M_{\rm He}$ at the onset of helium burning becomes smaller and smaller as $\alpha_{\rm ov,MS}$ increases.

6.1.3.2 Initial Chemical Composition

What is the impact of the initial chemical composition?

The influence of a different initial helium abundance is shown in Figure 6.8. A higher initial helium abundance increases the mean molecular weight, which in turn leads to a higher luminosity during MS (see Equation (3.11)). The MS convective core is thus larger and the impact of a higher He abundance is thus similar to that of adding an extra-mixed zone on top of the MS convective core, i.e., a lowering of the

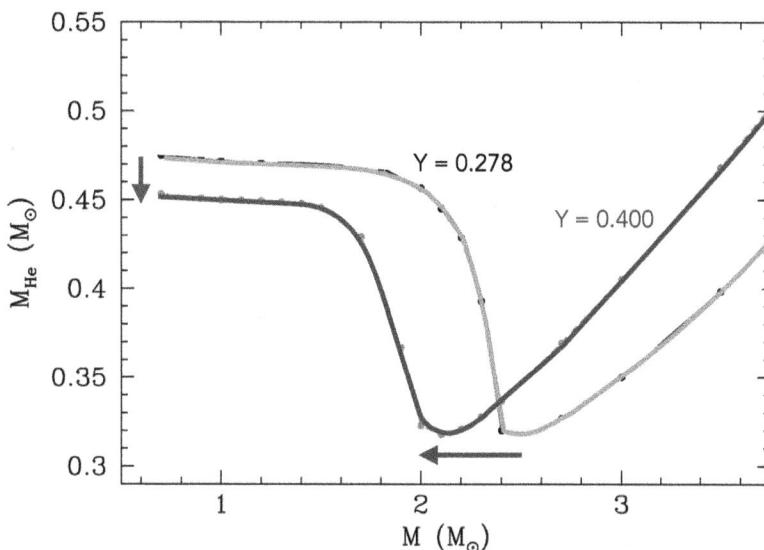

Figure 6.8. Mass of the helium core at the onset of helium burning as a function of the total stellar mass for different values of the initial helium abundance. Courtesy J. Montalbán.

transition mass. Moreover, Equation (5.12) shows that increasing Y (decreasing X) reduces the time interval necessary to reach the required helium burning temperature threshold. The increment in the helium core mass is thus smaller and M_{He} at the onset of the flash is smaller. This is especially true for large values of Y such as $Y = 0.4$ in Figure 6.8.

Figure 6.9 shows a similar graph for PARSEC models (Bressan et al. 2012) computed for a variety of metallicities and MS diffusive overshooting ($f_{ov,MS} = 0.025$). Decreasing the metallicity has a direct impact on the MS luminosity (see Equation (3.11)). Here again a higher MS luminosity makes the star behave like a slightly more massive star and the transition mass decreases. Furthermore, the mass of the degenerate helium core at the onset of the flash is all the larger as the metallicity is low. This can be explained by Equation (5.12), which shows that for a given mass of the helium core the increase in temperature with time is larger if X_{CNO}, i.e., the metallicity, is smaller. It thus takes more time to reach the threshold temperature to start helium burning and the H-shell also has more time to move outwards. This leads to a larger M_{He} at the onset of the flash.

6.1.4 Red Clump and Secondary Clump

What are the red and secondary clumps?

Figure 6.10 displays synthetic CMDs where accumulations of stars are clearly visible in the BV (top panels) and VI (bottom panels) planes (Girardi 1999). They

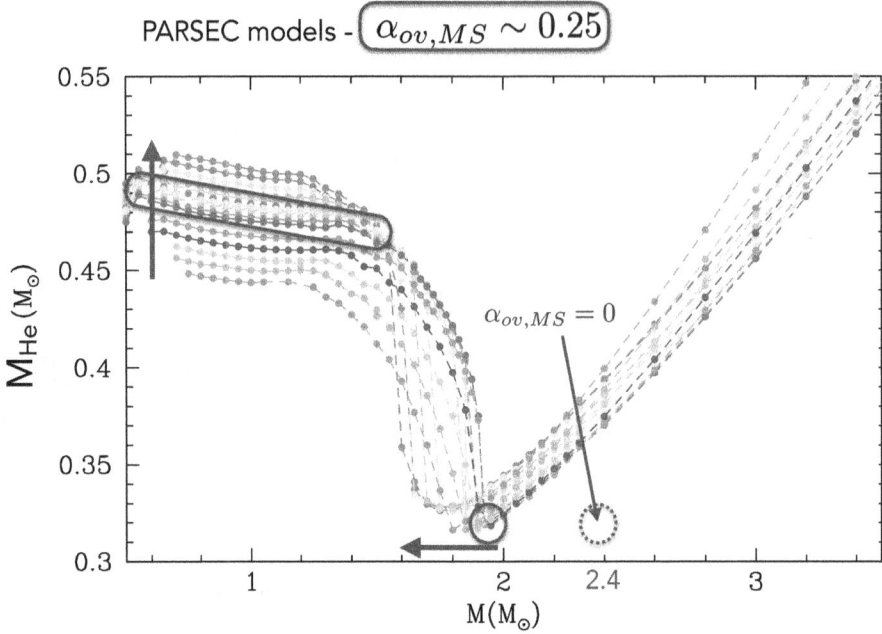

Figure 6.9. Same as Figure 6.4 for PARSEC models (Bressan et al. 2012) computed with MS diffusive overshooting ($f_{ov,MS} = 0.025$) and a variety of metallicities, from $Z = 5 \times 10^{-4}$ (bluest curve) to $Z = 7 \times 10^{-2}$ (reddest curve). Courtesy J. Montalbán.

show up as a central darker one and a lighter one located slightly below and on the left. The main one is known as the *Red Clump* (see Faulkner & Cannon 1973) while the less dense forms the *Secondary Clump* (see Girardi 1999). The location of these clumps depends on the metallicity, as can be seen in Figure 6.10.

How does the luminosity behave as a function of the stellar mass at the onset of helium burning?

Figure 6.11 displays the luminosity of core helium burning stars at the onset of helium burning as a function of the stellar mass. Let us first examine the insert, which is similar to Figure 6.4. It shows the mass of the helium core at the onset of helium burning as a function of the total stellar mass for two different assumptions for the initial helium abundance Y and a metallicity $Z = 0.02$. As can be expected from the discussion just above, the transition mass is smaller for the largest Y value while M_{He} for stars undergoing a flash is only slightly smaller.

Figure 6.11 also shows that the luminosity of helium burning models is not related to the sole mass of the He core as could be expected from Equation (5.9). Starting from the lowest stellar masses the luminosity increases in spite of a slightly decreasing helium core mass. This is due to the increased efficiency of the H-shell in a more and more massive envelope, which overcomes the decreased energy output of the helium burning core. Then, when the stellar mass increases, the decrease of the helium core mass eventually dominates and the luminosity reaches a minimum at the

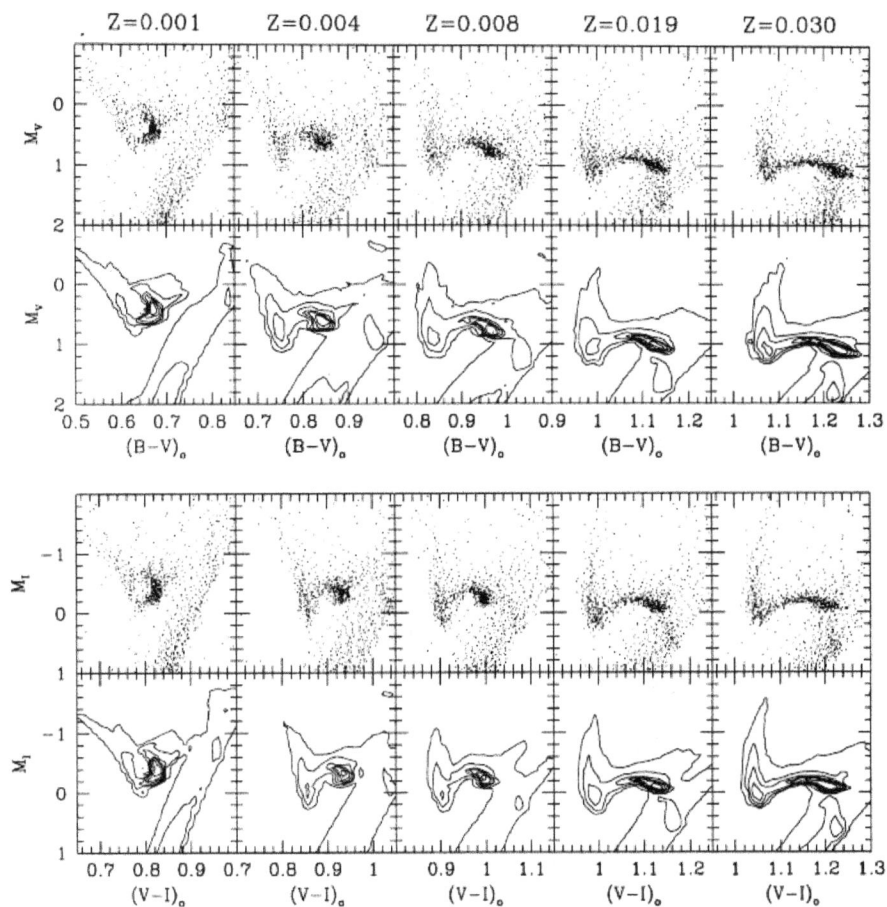

Figure 6.10. Synthetic CMDs showing the red clump region for different metallicities, in the BV (top panels) and VI (bottom panels) planes. Reproduced with permission from Girardi (1999).

transition mass. For stars more massive than the transition mass the increase in the helium core mass ($M_{He} \sim 0.1\ M$ see Section 6.1.2) induces a luminosity increasing with stellar mass.

Where are the red and secondary clumps in the HR diagram?

Figure 6.12 shows the location of Zero Age helium burning stars in the HR diagram with an insert reproducing Figure 6.11. Closely following the luminosity variation drawn in the inset we can see that all stars of masses in between 0.9 $M_\odot$ and the transition mass 2.4 $M_\odot$ are found in a very narrow location, which results in an accumulation of points in an observed HR diagram. This is the *Red Clump*. There is another accumulation coming from the fact that stars slightly below and slightly above the transition mass have very similar luminosities located in a very narrow V form. This is the *Secondary Clump*.

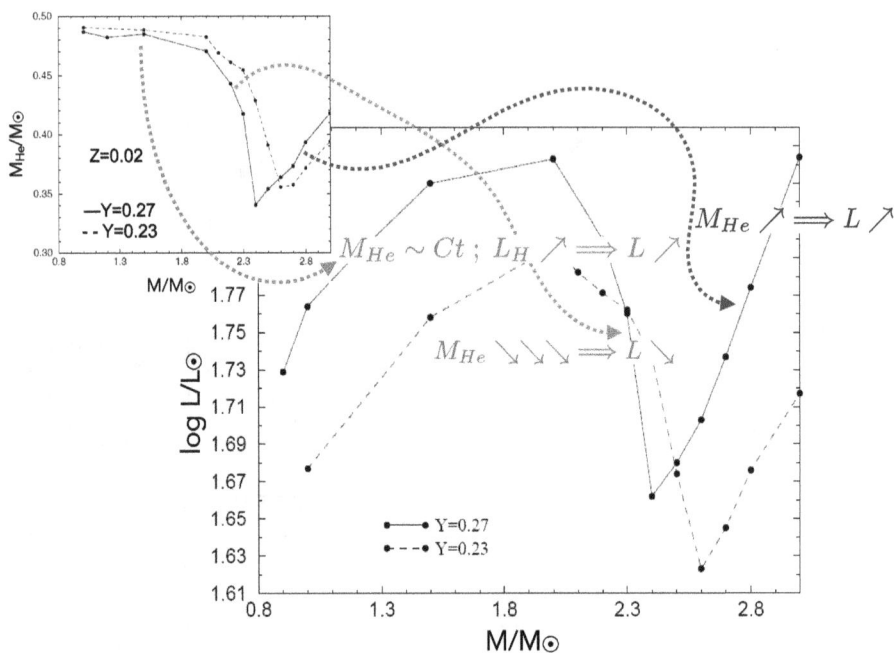

Figure 6.11. Luminosity of core helium burning stars at the onset of helium burning as a function of the stellar mass. The full and dotted curves have been computed with two different assumptions for the initial helium abundance and a solar metallicity ($Z = 0.02$). The inset shows a graph similar to Figure 6.4 with the same assumptions for Y and Z than the main graph. Reproduced with permission from Castellani et al. (2000).

Figure 6.13 displays the locations of the Red and Secondary Clumps in the HR diagram for models computed with CLES and AGSS09 abundances.

6.2 Quiescent Core Helium Burning

6.2.1 Convective Core

What is the extent of the convective core during core helium burning?

The presence of a convective core (see Section 3.3.2.1) is obvious in core helium burning stars since the temperature sensitivity is extremely high for the 3α reaction ($\sim$40, Equation (6.1)). There could however be a problem in locating the border of this convective core. As helium is burned into carbon, the opacity increases since the ionic absorption coefficient is proportional to Z^2. This increase in opacity bounces off on the radiative temperature gradient as can be seen in Figure 6.14 (Bossini et al. 2015). In the "Bare Schwarzschild" case (left panel), the boundary is found by a change of sign of $g = \nabla_{rad} - \nabla_{ad}$. It is clear that the condition for convective neutrality (see Section 2.1) is not met since ∇_{rad} becomes larger and larger at the boundary. On the contrary, when the boundary is found from an extrapolation of g (extrapolated Schwarzschild) from points located *inside* the convective core (right

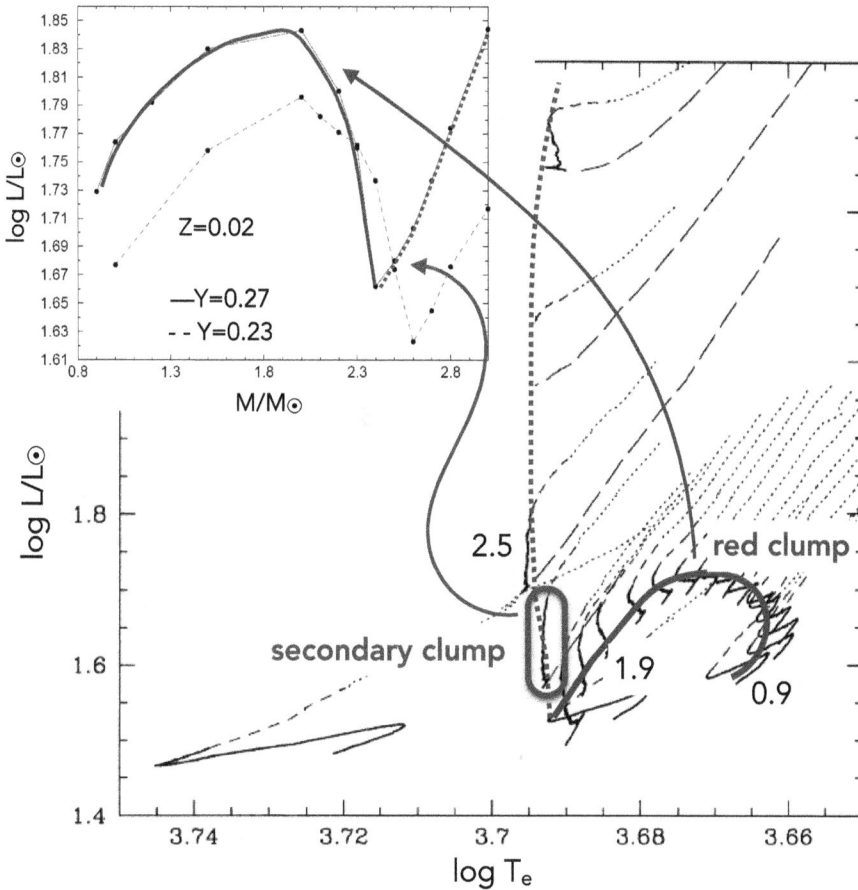

Figure 6.12. HR diagram of core helium burning stars showing the locations of the Red and Secondary Clumps. The inset is a copy of Figure 6.11. Reproduced with permission from Girardi (1999).

panel) the convective neutrality is perfectly fulfilled (see also Gabriel et al. 2014). This shows that the convective core grows in mass as helium burning proceeds.

The discontinuity in ∇_{rad} observed at the convective core boundary is due to the transition from a matter enriched in carbon to a "pure" helium matter. This means that at the convective core boundary, the discontinuity draws ∇_{rad} *below* ∇_{ad} in contrast to the discontinuity *above* ∇_{ad} found at the convective core boundary during MS (see Section 3.3.2.2).

6.2.2 Induced Semiconvection

What is an induced semiconvection?

Another difficulty was raised by Castellani et al. (1971) for more evolved models. This can be seen in Figure 6.15. Notwithstanding the temperature decrease with the

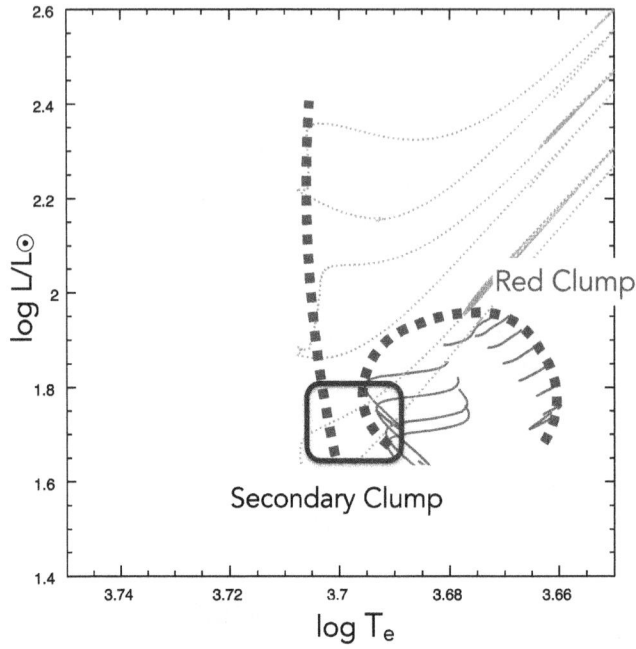

Figure 6.13. HR diagram of core helium burning stars showing the locations of the Red and Secondary Clumps for models computed with CLES.

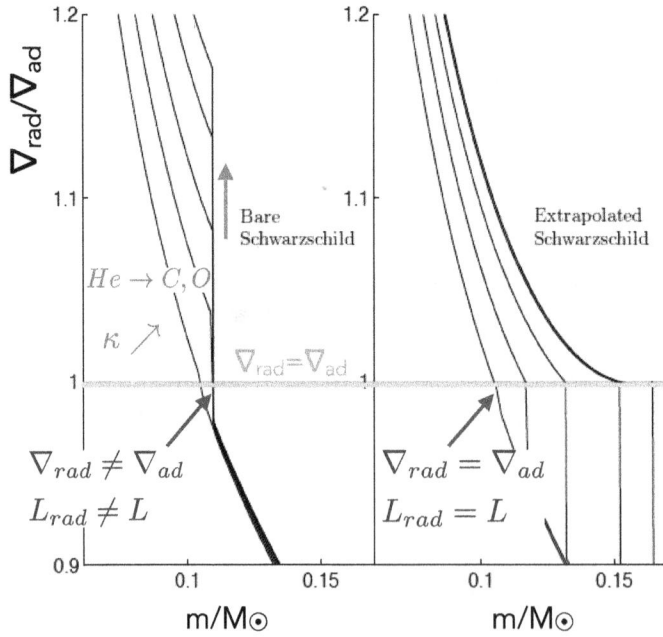

Figure 6.14. Temperature gradients ∇_{rad} and ∇_{ad} as a function of the fractional mass in the "bare Schwarzschild" and "extrapolated Schwarzschild" cases. Reproduced with permission from Bossini et al. (2015).

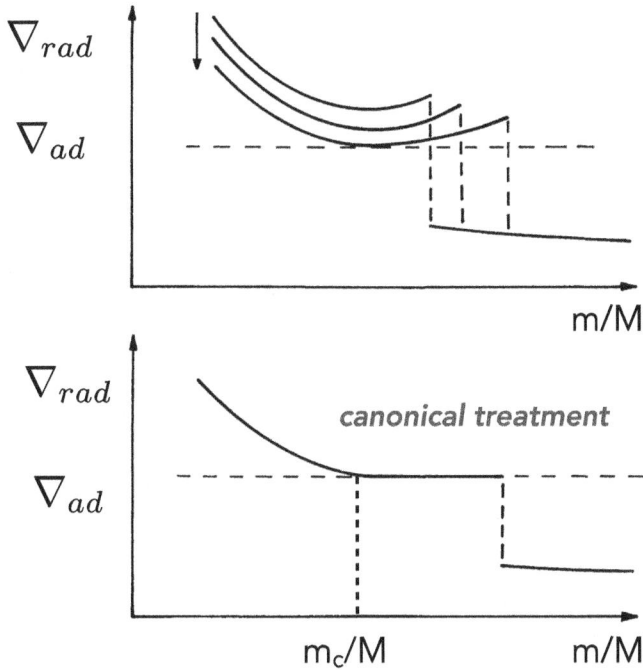

Figure 6.15. Temperature gradients ∇_{rad} and ∇_{ad} as a function of the fractional mass for more evolved models than in Figure 6.14. Reproduced from Castellani et al. (1971), with permission from Springer Nature.

fractional mass in the homogeneous helium burning core, a minimum appears in ∇_{rad}, which makes impossible a correct location of the convective core boundary. Decreasing the extent of the convective core increases g $(= \nabla_{rad} - \nabla_{ad})$ at the boundary while increasing it induces radiative layers in the "convective" core. A *canonical treatment* has been proposed by Castellani et al. (1971). It consists of moving the convective core boundary until the minimum in ∇_{rad} becomes exactly equal to ∇_{ad} and adapting the chemical composition of the overlying layers to ensure $g = 0$. This creates an *induced semiconvective* region whose outer boundary is discontinuous in chemical composition, density, opacity, and radiative gradient. This treatment is however quite tricky to implement in stellar evolution codes and most of the time a full or partial mixing in the semiconvective layers is adopted instead of the canonical treatment.

Such a canonical treatment has been adopted in CLES models. Figure 6.16 shows the temperature gradients ∇_{rad}, ∇_{ad}, and ∇_T in the inner 40% of the stellar mass for a star of 0.9 $M_\odot$. An overshooting parameter $\alpha = 0.2$ has been adopted (see Section 6.2.3). In the fully mixed overshooting region, the adopted temperature gradient ∇_T is the adiabatic gradient and the chemical composition is that of the convective core. In the semiconvective layers, the partial mixing is such that $\nabla_{rad} = \nabla_{ad}$. A downward discontinuity in temperature gradients and chemical composition is clearly seen.

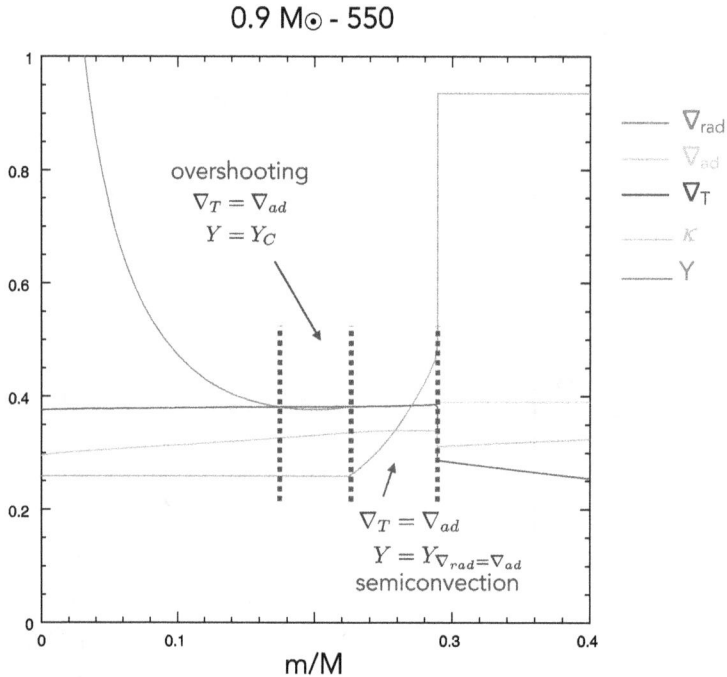

Figure 6.16. Temperature gradients ∇_{rad}, ∇_{ad}, and ∇_T as a function of the fractional mass (up to 0.4) for a CLES model of 0.9 $M_{\odot}$ computed with the canonical treatment of semiconvection.

Figure 6.17 shows the behavior of the convective core mass, the overshooting region, and the semiconvective zone in a star of 0.9 $M_{\odot}$ computed with CLES during core helium burning. Interestingly, the boundary of the convective core and that of the overshooting region remain locked while the semiconvective boundary grows up to the end of core helium burning.

Figure 6.18 is adapted from an animation, showing the evolution with time of the temperature gradients ∇_{rad}, ∇_{ad}, and ∇_T. The reader should be attentive to the changing helium profile in the semiconvective layers. The chemical composition is indeed imposed to fulfill the condition $\nabla_{rad} = \nabla_{ad}$.

6.2.3 Extra-Mixing

What is the impact of extra mixing?

The mixed or partially mixed region can even be intensified by the addition of an extra mixing above the convective core. An obvious effect of such an extra mixing is found in the core helium burning lifetime, which can be drastically increased when adopting quite extreme assumptions as to the amount of additional mixing (see Section 3.3.2.3). Figure 6.19 shows the extent of the convective core (upper panel)

0.9 M⊙

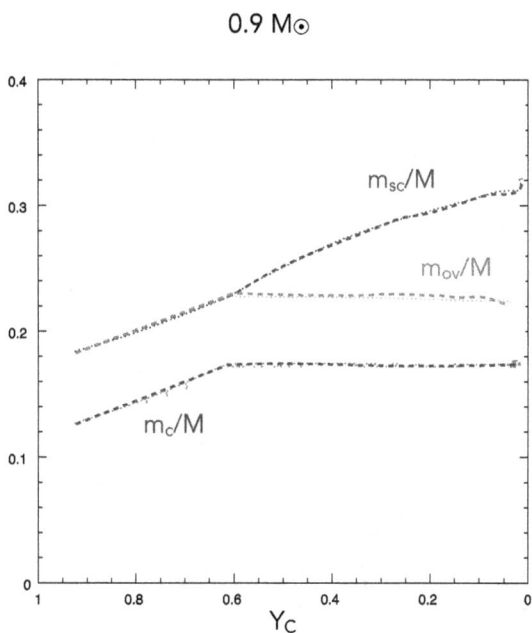

Figure 6.17. Evolution of the convective core mass, the overshooting region, and the semiconvective zone in a star of 0.9 $M_\odot$ computed with CLES as a function of the central helium abundance.

0.9 M⊙

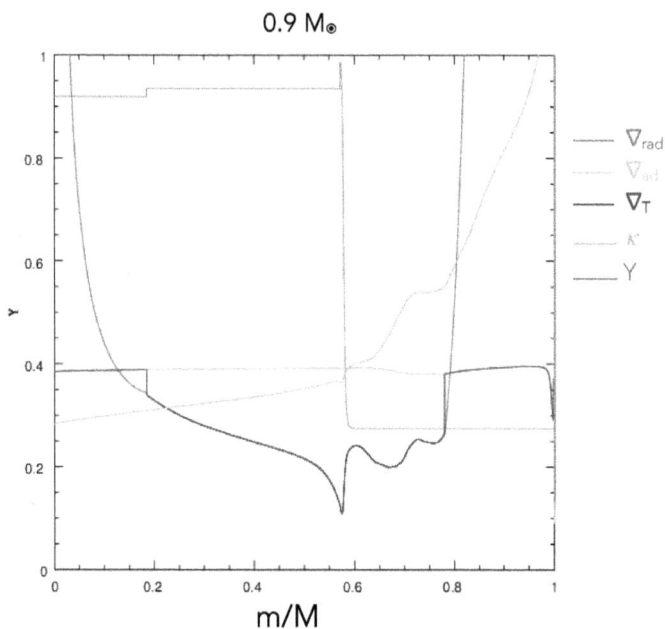

Figure 6.18. Figure showing the evolution of the convective core, the overshooting region and the semi-convective zone during core helium burning. The temperature gradients ∇_{rad}, ∇_{ad}, ∇_T are shown in red, cyan, and blue, respectively. The helium abundance, Y, and the opacity, κ, are, respectively, displayed in green and light brown. This figure corresponds to Movie6 in the supplementary Keynote and pptx files, accessible at http://doi.org/10.1088/2514-3433/adcf15.

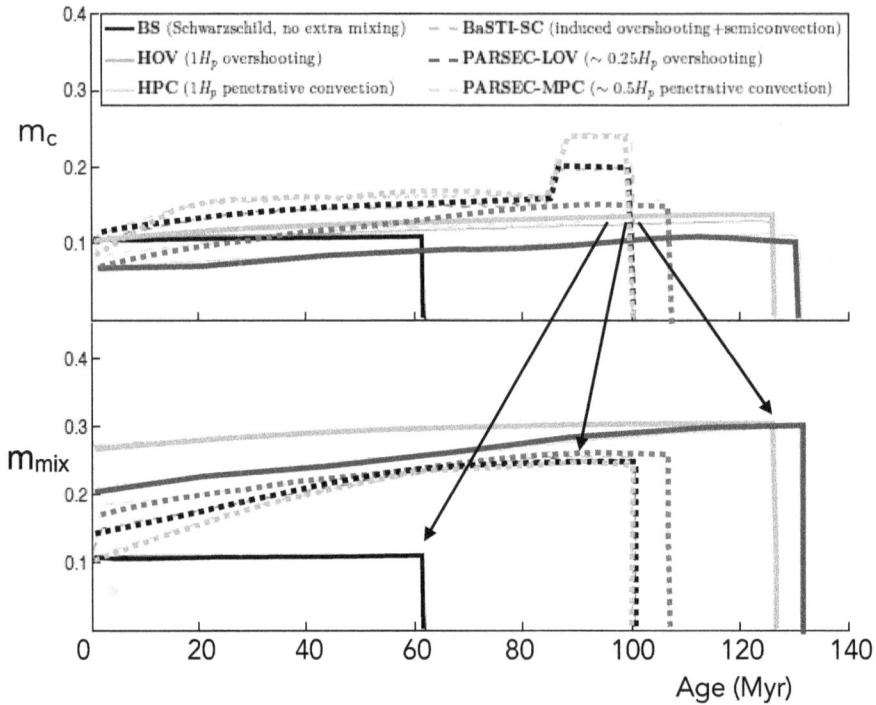

Figure 6.19. Extent of the convective core (upper panel) and of the mixed region (lower panel) during core helium burning as a function of time in models computed with different assumptions listed in the inserted legend. Reproduced with permission from Bossini et al. (2015).

and of the mixed region (lower panel) based on different assumptions for the mixing listed in the inserted legend (see Bossini et al. 2015 for details on the assumed mixing processes). Another important impact of extra mixing is the longer time available to the H-shell to move outwards, which leads to a larger helium core at the end of core helium burning.

6.2.4 Horizontal Branch

What does "horizontal branch" mean?

Depending on the amount of hydrogen-rich matter on top of the H-shell, the effective temperature of core helium burning stars is different, the smaller the envelope mass, the larger the effective temperature. In a GC color–magnitude diagram, these helium burning stars form the *Horizontal Branch* discovered by Arp et al. (1952). An example of the morphology of the horizontal branch is given in Figure 6.20 for the GC NGC 2808 (Gaia Collaboration et al. 2016). This morphology is an extremely complex problem, which will not be addressed here.

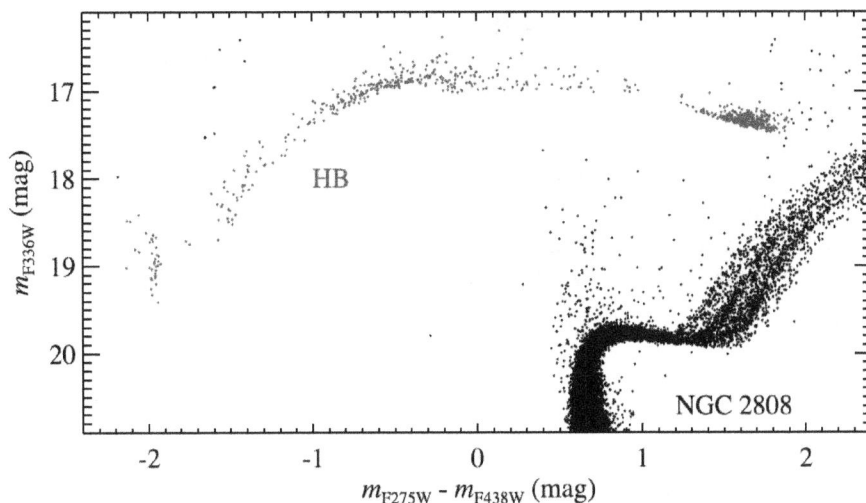

Figure 6.20. Color–magnitude diagram of the GC NGC 2808. Blue points are horizontal branch stars. Credit: Gaia Collaboration et al. (2016), reproduced with permission © ESO.

6.3 List of Questions

Why is the temperature increasing in a contracting degenerate core?
What is the mass of the helium core at the onset of helium burning in low mass stars?
Why is there a flash at the onset of helium burning?
What are the core temperature and density behaviors during the flash?
What is the behavior of an intermediate mass star?
What is the mass of the helium core at the onset of core helium burning?
What can affect the transition mass?
What is the impact of MS extra mixing?
What is the impact of the initial chemical composition?
What are the red and secondary clumps?
How does the luminosity behave as a function of the stellar mass at the onset of helium burning?
Where are the red and secondary clumps in the HR diagram?
What is the extent of the convective core during core helium burning?
What is an induced semiconvection?
What is the impact of extra mixing?
What does "horizontal branch" mean?

References

Arp, H. C., Baum, W. A., & Sandage, A. R. 1952, AJ, 57, 4
Bossini, D., Miglio, A., Salaris, M., et al. 2015, MNRAS, 453, 2290
Bressan, A., Marigo, P., Girardi, L., et al. 2012, MNRAS, 427, 127
Castellani, V., Degl'Innocenti, S., Girardi, L., et al. 2000, A&A, 354, 150
Castellani, V., Giannone, P., & Renzini, A. 1971, Ap&SS, 10, 355

Faulkner, D. J., & Cannon, R. D. 1973, ApJ, 180, 435

Gabriel, M., Noels, A., Montalbán, J., & Miglio, A. 2014, A&A, 569, A63

Gaia Collaboration, Brown, A. G. A., Vallenari, A., et al. 2016, A&A, 595, A2

Girardi, L. 1999, MNRAS, 308, 818

Kippenhahn, R., & Weigert, A. 1994, Stellar Structure and Evolution (Berlin: Springer)

Montalbán, J., & Noels, A. 2013, EPJ Web Conf., 43, 03002

Salaris, M., & Cassisi, S. 2006, Evolution of Stars and Stellar Populations (New York: Wiley)

Schinder, P. J., Schramm, D. N., Wiita, P. J., Margolis, S. H., & Tubbs, D. L. 1987, ApJ, 313, 531

Ventura, P., D'Antona, F., & Mazzitelli, I. 2008, Ap&SS, 316, 93

Chapter 7

Asymptotic Giant Branch

7.1 Early AGB Phase (EAGB Phase)

Where does the formation of a He-burning shell take place?

7.1.1 Formation of a Helium Burning Shell (He-Shell)

In a way similar to what happened at the TAMS the exhaustion of helium in the core gives rise to the formation of a helium burning shell (He-shell). The location of this He-shell depends on the extent of the mixed region during core helium burning and thus is extremely sensitive to the assumptions made in this regard. The luminosity, nearly constant during core helium burning, starts again to increase and for the second time our star climbs the Hayashi track. The structure is very similar to the earlier RGB structure and the evolutionary path is a sort of extension of the RGB although not really superimposed. It is indeed "asymptotically" close to the RGB, which is at the origin of its name: *Asymptotic Giant Branch.*

7.1.2 AGB Bump and Second Dredge-Up

What is the physical origin of the AGB bump?

Figure 7.1 shows a feature that is quite similar to the RGB bump (Section 5.3) since it shows a temporary decrease in luminosity followed by a new rise along the Hayashi track (observationally discovered by Gallart (1998)). The physical origin of this bump is related to the development of the He-shell, which requires an expansion and cooling of the outer layers in order to maintain the strict temperature conditions within the He-shell. This first produces a near extinction of the H-shell and a decrease in the efficiency of the He-shell. A global contraction ensues with a decrease in luminosity (see Catelan 2007 and references therein). With the stabilization of the newly formed He-shell, the luminosity resumes its increasing behavior along the AGB track.

7-1

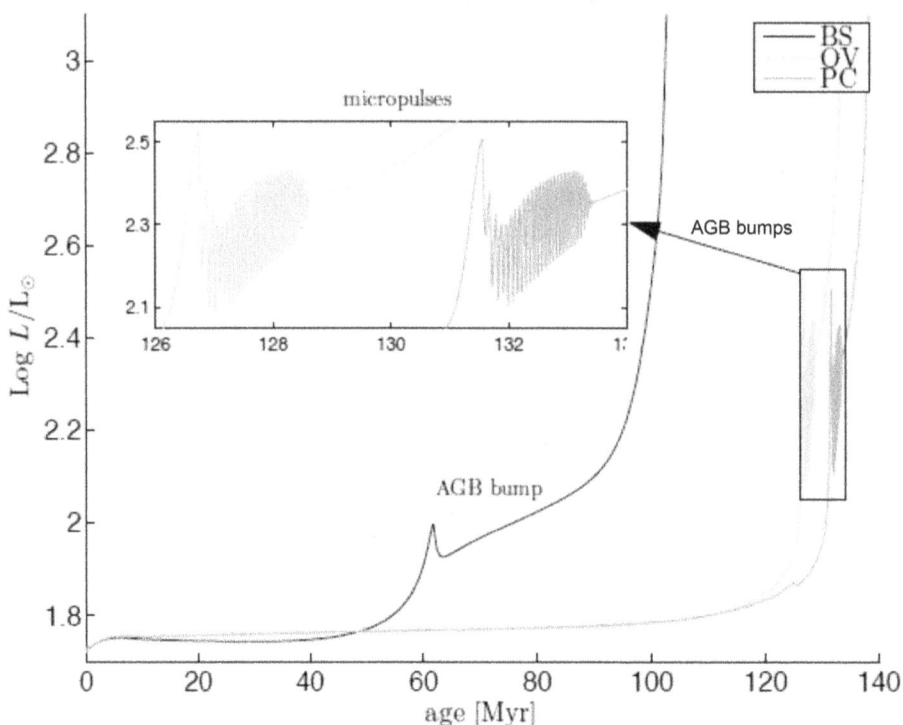

Figure 7.1. Evolution of the luminosity with time in a core helium burning and He-shell burning 1.5 $M_\odot$ star computed with different mixing assumptions. The inserted legend refers to the legend in Figure 6.19. Reproduced with permission from Bossini et al. (2015).

Figure 7.1 also shows that the luminosity of the bump is larger when extra-mixing is taken into account during core helium burning. This is not surprising since, as we have seen in Section 6.2.3, extra-mixing not only increases the CO core but also the helium core and the luminosity is thus larger at the formation of the He-shell (see Section 5.2).

7.1.2.1 Micropulses

What is the driving mechanism of micropulses?

After the maximum luminosity of the bump, a thermal instability may be encountered if nuclear reactions in the core have stopped significantly before the formation of the He-shell, which happens if the centrally mixed region is large enough during core helium burning. With an extended CO core, the physical conditions to start helium burning on top of the CO core are not immediately fulfilled and a more important contraction of the core is required. This produces a He-shell, which presents the *"thinness"* characteristics inherent to a secular instability first described by Schwarzschild & Härm (1965) in the context of thermally pulsating AGB stars (see Section 7.2), i.e.,

$$\Delta T/T > 4/\nu \tag{7.1}$$

$$\Delta r/r < 5/(2|Q|) \tag{7.2}$$

where ν is the temperature sensitivity of the nuclear reactions ($\nu \sim 38 - 40$) and Q has a typical value of -6.

The He-shell must indeed be "thin" enough in radius (Equation (7.1)) and "thick" enough in temperature (Equation (7.2)) to maintain hydrostatic equilibrium even in the presence of a thermal runaway. Despite a drastic temperature difference between the He-shell and the adjoining layers, the pressure within the shell is on the whole maintained and a cooling by expansion cannot stabilize the shell. Even though matter is non-degenerate in the He-shell, the thermostatic pressure control mechanism (see Section 3.2.1) does not act because of the thinness of the shell and the thermal conditions are similar to those of the He flash (see Sections 4.3 and 6.1.1).

This thermal runaway shows up as a series of recurrent bumps (see the inset in Figure 7.1), which were discovered by Mazzitelli & Dantona (1986) who introduced the notion of *micropulses* (see also Gautschy & Althaus 2007). During the heating phase of the micropulse, the thinness conditions become less fulfilled since this heating leads to an enlargement of the shell. The thermostatic pressure control starts to act and a cooling and narrowing of the shell ensues. This in turn renders its thinness conditions to the shell and a new micropulse is fired up. After each micropulse, the physical conditions slightly change and the thinness aspect is eventually lifted, which marks the end of the series of micropulses. They are not present with the bare Schwarzschild assumption since they need a larger amount of mixing during core helium burning to fully develop and a simple AGB bump is seen in Figure 7.1.

Figure 7.2 shows micropulses for different stellar masses computed with CLES in a $\log T_c$, $\log \rho_c$ diagram. Models of our favorite 1.3 $M_\odot$ displayed in the left panel have been obtained from a post-flash model.

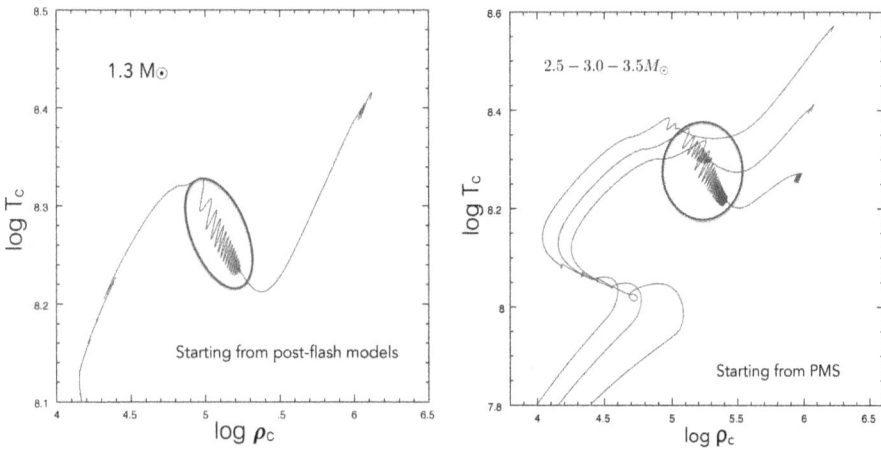

Figure 7.2. Micropulses in a $\log T_c$, $\log \rho_c$ diagram for a 1.3 $M_\odot$ star starting from a post-flash model (left panel) and for stellar masses of 2.5, 3.0, and 3.5 $M_\odot$ (right panel) computed with CLES.

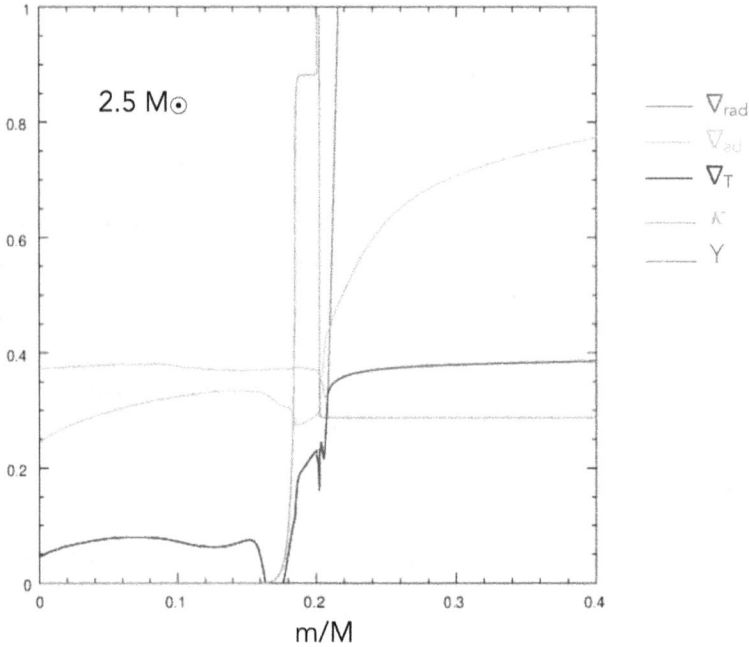

Figure 7.3. Figure showing the temperature gradients ∇_{rad}, ∇_{ad}, and ∇_T (in red, cyan, and blue, respectively) during micropulses in a star of 2.5 $M_\odot$. The helium abundance and the opacity are displayed in green and light brown, respectively. This figure corresponds to Movie7 in the supplementary Keynote and pptx files, accessible at http://doi.org/10.1088/2514-3433/adcf15.

The evolution of the temperature gradients ∇_{rad}, ∇_{ad}, and ∇_T during micropulses in a star of 2.5 $M_\odot$ is shown in Figure 7.3. The reader can follow the behavior of:

- the conditions within the shell with the drastic increase of the radiative temperature gradient during the flash phase of the micropulse, which leads to the ocurence of convection in the shell,
- the convective envelope through ∇_{rad} (red curve) outside the He-shell, which sort of breathes with expansion and recess as the flash develops and then calms down.

7.1.3 AGB Bump Constraint on Extra-Mixing

What can we learn from the location of the AGB bump?

Although much less apparent than the Red Clump the AGB bump can be visualized in Figure 7.4, which shows a luminosity histogram from the APOKASC catalog (Pinsonneault et al. 2014) (let panel) and from a TRILEGAL (Girardi et al. 2005) simulation of the galactic population in the Kepler field (Bossini et al. 2015).

The location and the detailed features of the AGB bump bring an important constraint on the extent of extra-mixing during core helium burning. Figure 7.5 shows the comparison of the distribution of stars in the APOKASC catalog

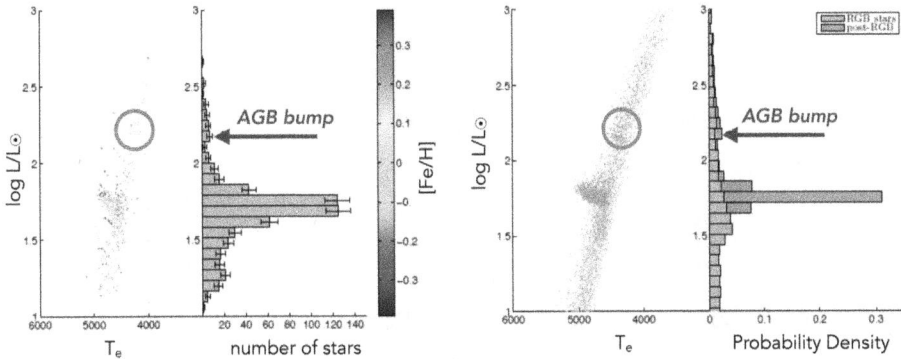

Figure 7.4. Location of the AGB bump in a luminosity histogram from the APOKASC catalog (Pinsonneault et al. 2014) (let panel) and from a TRILEGAL (Girardi et al. 2005) simulation of the galactic population in the Kepler field. Reproduced with permission from Bossini et al. (2015).

(Pinsonneault et al. 2014) and in TRILEGAL (Girardi et al. 2005) simulations computed with different assumptions regarding the extra-mixing during core helium burning (Bossini et al. 2015). It is clear that the bare Schwarzschild, with a too small (wrong) convective core, must be rejected as well as assumptions leading to a too extended extra-mixed core (upper panel). Intermediate assumptions are clearly favored (bottom panel).

7.1.4 Early ABG Phase

After the micropulse phase, the evolution quietly proceeds along the AGB and the stellar structure is as follows:

- an inert and more and more degenerate carbon oxygen core (CO core), which mass is slowly increasing,
- a thin non-degenerate very active He-shell located at the border of the CO core and still embedded in "pure" helium layers,
- a thin "sleeping" non-degenerate H-shell located at the H/He discontinuity, and
- an expanding hydrogen-rich envelope.

As long as the He-shell moves outwards in a helium-rich matter still far enough below the H-shell, the evolutionary phase is named *Early AGB* phase.

With the "sleeping" behavior of the H-shell, the convective envelope is allowed to penetrate into deeper layers. In stars more massive than about 4 $M_\odot$, the envelope eventually reaches layers whose chemical composition has been affected by hydrogen CNO burning. These products are brought to the surface through a *second dredge-up* with an enrichment in He^4 and N^{14} and a depletion in C^{12} and O^{16}. This second dredge-up does not occur in lower mass stars.

The sleeping H-shell is located at the boundary of the helium core with a discontinuity in H/He. The He-shell progressively moves outwards and eventually reaches this discontinuity. With a sudden lack of helium, the He-shell drastically

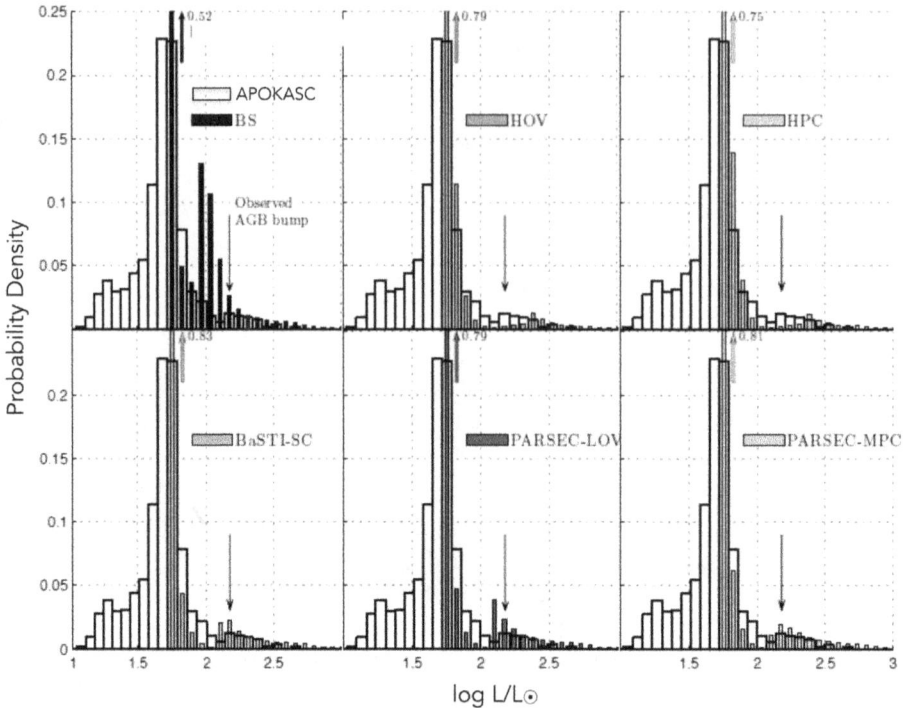

Figure 7.5. Comparison of the distribution of stars in the APOKASC catalog (Pinsonneault et al. 2014) and in TRILEGAL (Girardi et al. 2005) simulations computed with different assumptions regarding the extra-mixing during core helium burning. Reproduced with permission from Bossini et al. (2015).

fades out and a contraction of the overlying layers occurs. This reignites the H-shell and signs the end of the Early AGB phase and the start of the Thermally Pulsating AGB phase.

7.2 Thermally Pulsating AGB Phase (TP-AGB Phase)

What is the physical origin of the AGB pulses?

7.2.1 Pulse and Interpulse

Our favorite star is now reaching the onset of a rather complex evolution which translates into a series of thermal pulses. This is the *Thermally Pulsating AGB* phase (TP-AGB). The evolution proceeds through highly unstable interactions between the He-shell and the H-shell, which are indeed unable to gently cohabit. Each time one shell dominates the energy production, it is done at the expense of the other. This translates into a series of thermal pulses with an increasing mean luminosity (see Salaris & Cassisi 2006 for more details).

Such a pulse is illustrated in Figure 7.6 where the energy outputs from the H-shell and He-shell are, respectively, drawn in green and red, while the surface radius and

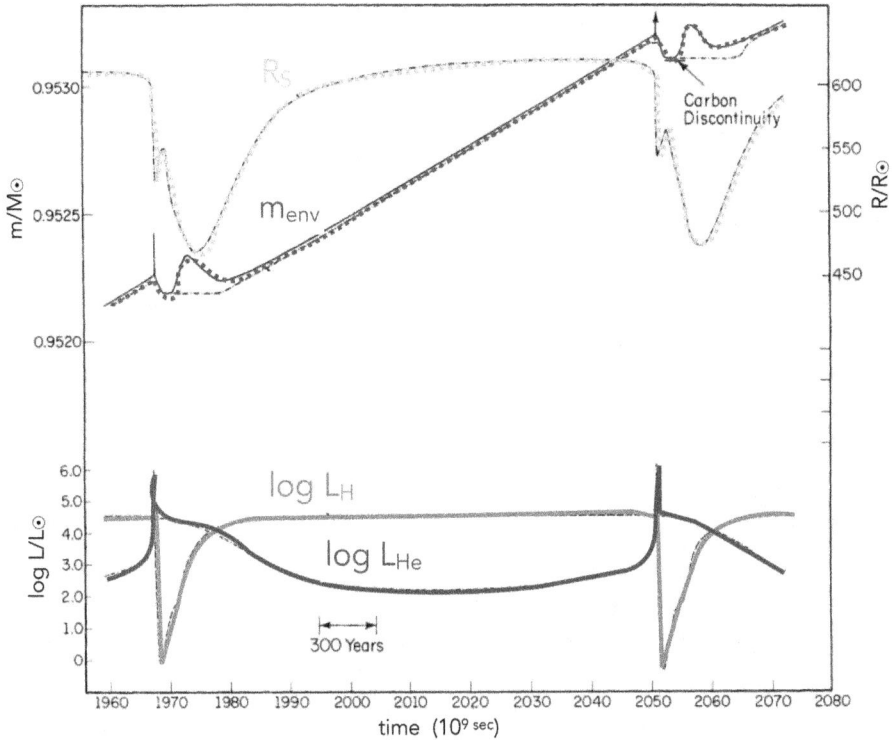

Figure 7.6. Thermal pulse in a 5 $M_\odot$ star as a function of time. The energy outputs from the H-shell and He-shell are, respectively, drawn in green and red. The surface radius and the mass at the bottom of the convective envelope are displayed by dotted cyan and blue curves. Reproduced from Iben (1975). © 1975. The American Astronomical Society. All rights reserved. Printed in U.S.A.

the mass at the bottom of the convective envelope are shown in dotted cyan and blue curves (adapted from Iben 1975).

In between two consecutive pulses, a quiet *interpulse* phase is seen. Most of the interpulse is fueled by the energy produced by the H-shell and the convective envelope recedes and expands as the H-shell moves outward. However, this displacement of the H-shell imposes a heating of the adjoining layers, which eventually reignites the He-shell. As was the case at the start of the micropulses (Section 7.1.2), the conditions of "thinness" are satisfied and this reignition is secularly unstable and takes the form of a flash. The thermostatic pressure control mechanism is impeded, which leads to a thermal runaway with a sudden increase of the energy produced by the He-shell. Such a heating is accompanied by an enlargement of the He-shell, which progressively loses its thinness characteristics and allows the thermostatic pressure control mechanism to operate. However, the resulting expansion and cooling of the He-shell and adjoining layers induces the switch-off of the H-shell. A global contraction ensues and the H-shell is accordingly reignited, signing the end of the flash and the start of a new interpulse quiet phase.

Figure 7.7 shows thermal pulses for different stellar masses computed with CLES in a log T_c, log ρ_c diagram.

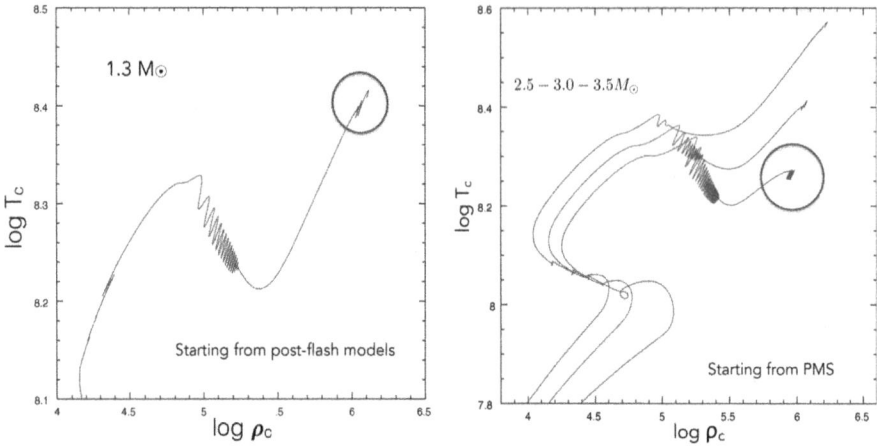

Figure 7.7. Thermal pulses in a log T_c, log ρ_c diagram for a 1.3 $M_\odot$ star starting with a post-flash model (left panel) and for stellar masses of 2.5, 3.0, and 3.5 $M_\odot$ (right panel) computed with CLES.

The evolution of the temperature gradients ∇_{rad}, ∇_{ad}, ∇_T during thermal pulses in a star of 2.5 $M_\odot$ is shown in Figure 7.8. The physical conditions within the shell are similar to those observed during micropulses. The reader can follow the behavior of:

- the conditions within the shell with the drastic increase of the radiative temperature gradient during the flash phase of the micropulse, which leads to the occurrence of convection in the He-shell,
- the convective envelope through ∇_{rad} (red curve) outside the He-shell, which sort of breathes with expansion and recess as the flash develops and then calms down.

7.2.2 Third Dredge-Up

What are the origin and the impact of the third dredge-up?

During the most active phase of the pulse, the energy output is such that a pulse-driven convective zone (PDCZ) is formed in the area of the He-shell and the matter processed by helium burning is spread out on layers located around the most active part of the shell. Figure 7.9 shows in an À la Kippenhahn diagram the evolution of the PDCZ in a TP-AGB star during and between two consecutive thermal pulses (Marigo et al. 2013). The convective envelope is shown in violet, while the CO core is displayed in light blue. The PDCZ, drawn in orange with a blue contour, is not present during the interpulse phase when energy is essentially produced by the H-shell. This leads to a stratification of chemicals shown in the right part of Figure 7.9 just before the onset of the new PDCZ with, from top to bottom, the ashes from the H-burning shell (top—green), from the previous PDCZ (middle—orange), and from the He-burning shell (bottom—blue). Once convection sets in, these ashes are mixed producing a chemical mixture, which is hydrogen-rich but also carbon and helium enriched.

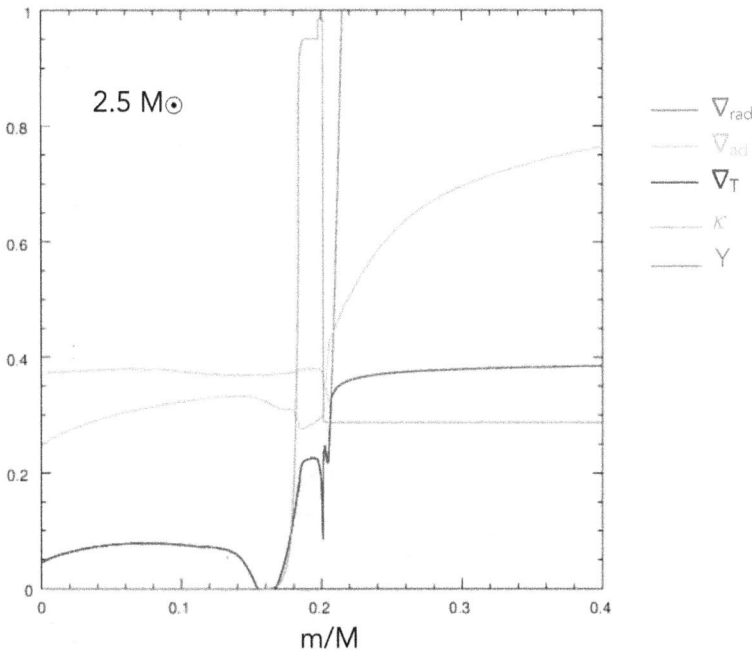

Figure 7.8. Figure showing the temperature gradients ∇_{rad}, ∇_{ad}, ∇_T (in red, cyan, and blue, respectively) during thermal pulses in a star of 2.5 $M_\odot$. The helium abundance and the opacity are displayed in green and light brown, respectively. This figure corresponds to Movie8 in the supplementary Keynote and pptx files, accessible at http://doi.org/10.1088/2514-3433/adcf15.

As can be seen in Figures 7.6 and 7.9, the convective envelope is somewhat pushed back during the quiet phase of H-shell burning, but during the cooling phase of the pulse, the convective envelope can resume a brief penetration in matter enriched in C and O. This is at the origin of the *third dredge-up*.[1] This dredge-up of helium burning processed material, in particular C and O, is at the origin of carbon- and oxygen-rich AGB stars.

7.2.3 S-Process Nucleosynthesis

The third dredge-up has a tremendous importance in the processing of heavy elements through neutron captures. At the reignition of the H-shell in a matter enriched in helium, neutrons are produced through the reaction

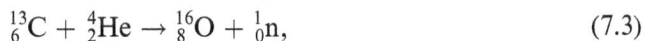

$$^{13}_{6}\text{C} + {}^{4}_{2}\text{He} \rightarrow {}^{16}_{8}\text{O} + {}^{1}_{0}\text{n}, \tag{7.3}$$

or, in a less efficient way, by the reaction

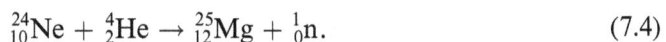

$$^{24}_{10}\text{Ne} + {}^{4}_{2}\text{He} \rightarrow {}^{25}_{12}\text{Mg} + {}^{1}_{0}\text{n}. \tag{7.4}$$

---

[1] Interestingly enough stars with mass lower than $\sim$4 $M_\odot$ have a first dredge-up and a third dredge-up but not a second dredge-up (see Section 7.1.2).

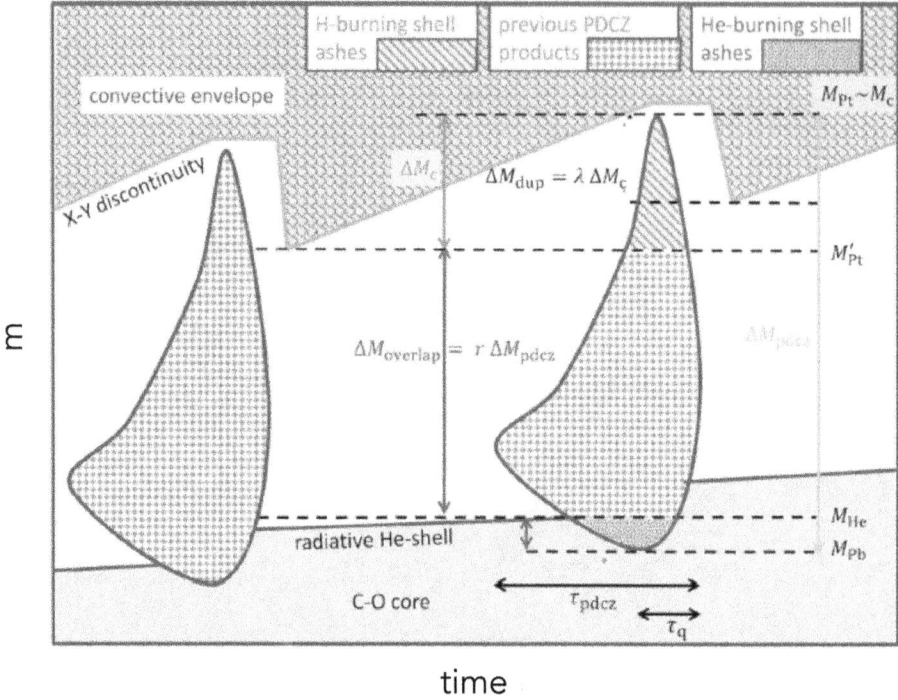

Figure 7.9. À la Kippenhahn diagram showing the evolution of the inner layers in the two-shell structure of a TP-AGB star during and between two consecutive thermal pulses. The PDCZ is drawn in orange with a blue contour. The right part of the figure shows the three-zone stratification of the material just before the development of the convective pulse, containing from top to bottom: the ashes left by the H-burning shell, the products of the previous PDCZ, and the ashes left by the He-burning shell. Reproduced with permission from Marigo et al. (2013).

These neutron sources are indeed the driving force of the s-process nucleosynthesis[2] (for an early review on nucleosynthesis, see Clayton 1969). Neutrons are captured by the *in situ* elements to form heavier isotopes beyond Sr and Y and up to Pb.

The successive penetrations of the envelope brings s-process elements to the surface with abundances strongly related to the recurrence of the mixing in the PDCZ. Clayton (1968) has showed that this recurrence was indeed the necessary condition to reproduce the observed abundances of s-process elements.

After the most active part of each pulse and the spread resulting from the short-lived presence of the PDCZ, the "thinness" conditions of a secular instability are removed (see Equation (7.1)) and, as discussed in Section 7.2.1, conditions for a new interpulse are met.

Figure 7.10 shows the changes in the surface metallicity for a set of isochrones with ages varying from $\sim 10^8$ to $\sim 10^{10}$ years and a color coded metallicity (Marigo et al. 2017). The reduced metallicity during MS for low mass stars due to diffusion is

---

[2] The s in s-process means *slow* as opposed to r in r-process referring to *rapid*. This r-process is also a neutron capture process mainly occurring during the explosive phase of a type II supernova, on a timescale of a few seconds only.

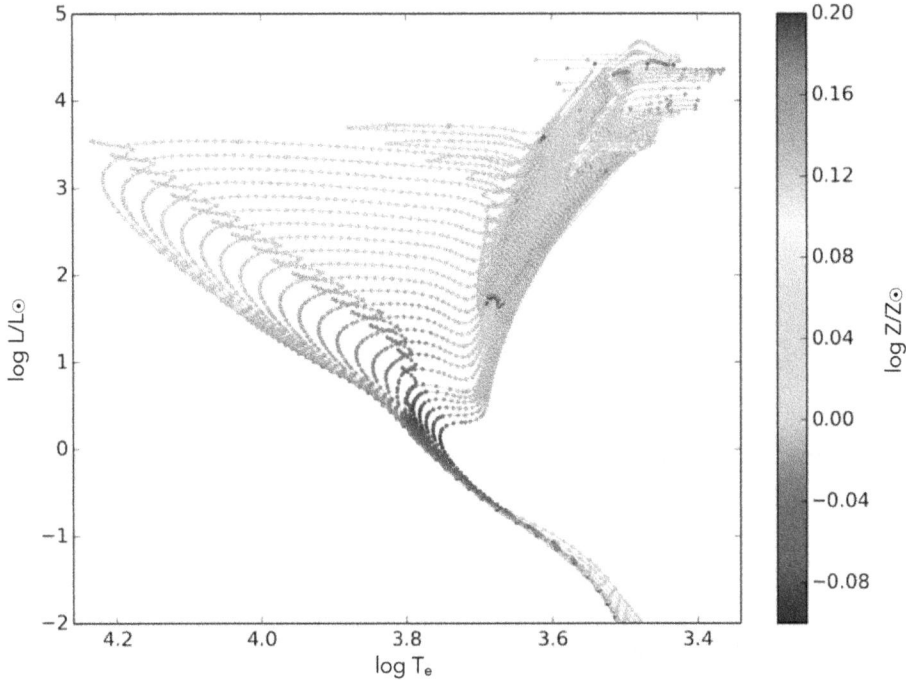

Figure 7.10. Changes in the current surface metal content Z across the HR diagram, for a set of isochrones of initial $Z = 0.01471$, with ages varying from $\log(t \text{ years}) = 7.8–10.1$ at steps of 0.1 dex. Reproduced from Marigo et al. (2017). © The American Astronomical Society. All rights reserved.

clearly visible in blue, while the drastic increase of about 60% resulting from the third dredge-up appears in yellow to red at high luminosity on the AGB. A tiny signature of the second dredge-up in intermediate mass stars is also visible while the first dredge-up is almost imperceptible along the RGB.

7.2.4 Luminosity at the Tip of the AGB

Why is there a maximum AGB luminosity?

Even though the luminosity goes up and down during the pulses, following the envelope expansions and contractions, its mean value increases. There is however a limit to the increase in luminosity. The *Eddington luminosity* is the maximum luminosity a star can achieve when the gravitational force acting inward is exactly balanced by the force resulting from the gradient of radiation pressure acting outward.

Hydrostatic equilibrium in the layers close to the surface imposes that

$$\frac{\partial P}{\partial r} = -\frac{G\rho M}{R^2}. \tag{7.5}$$

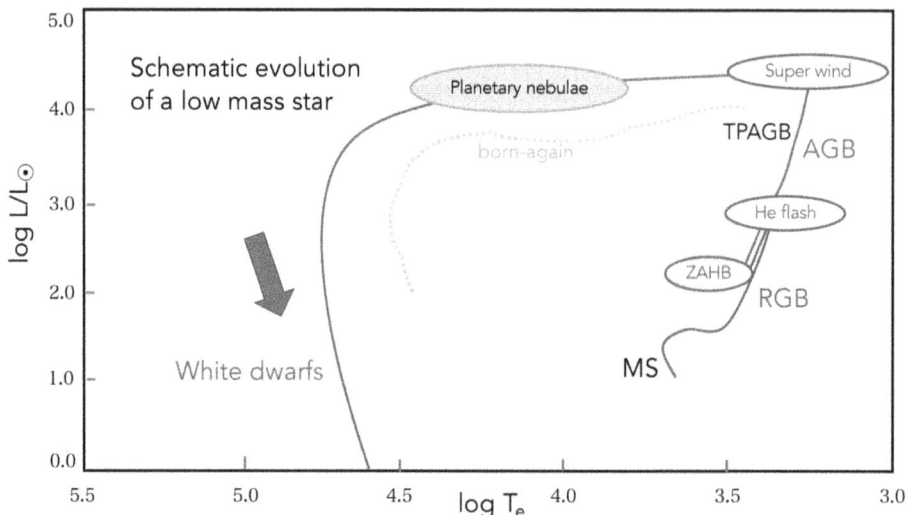

Figure 7.11. Evolutionary track for a 2 $M_{\odot}$ star from the ZAMS to the white dwarf cooling phase. The blue track is a born-again event shifted in L and T_e for clarity. Adapted from Herwig (2005).

If the pressure is essentially dominated by radiation pressure, we can write

$$\frac{\partial P}{\partial r} = \frac{4}{3}\, aT^3 \frac{\partial T}{\partial r} = -\frac{\kappa \rho L}{4\pi c R^2} \tag{7.6}$$

where a is the radiation constant. Equating these two relations gives

$$L_{\text{Edd}} = \frac{4\,\pi c G M}{\kappa} \sim 3.3 \times 10^4 \frac{M}{M_{\odot}} L_{\odot} \tag{7.7}$$

which is the Eddington limit to the luminosity allowed for a star in hydrostatic equilibrium. Above that limit a radiation driven extremely strong wind supposedly ejects the outer stellar layers. This is, for instance, what is thought to be experienced by massive stars reaching the phase of luminous blue variables (LBV).

All along the AGB, stars are subject to "ordinary" mass loss through stellar wind. This mass loss is followed by a *super-wind* with an extremely large mass loss rate. This super-wind could possibly be triggered by pulsations occurring at the tip of the AGB (see, for instance, McDonald & De Beck 2018 and references therein) and is probably at the origin of long period variables such as Mira stars. Unfortunately the physical details of these events, as well as those of the whole AGB phase, are still a matter of debate (see Lattanzio & Karakas 2016) (Figure 7.11).

7.3 List of Questions

Where does the formation of a He-burning shell take place?
What is the physical origin of the AGB bump?
What is the driving mechanism of micropulses?
What can we learn from the location of the AGB bump?

What is the physical origin of the ABG pulses?
What are the origin and the impact of the third dredge-up?
Why is there a maximum AGB luminosity?

References

Bossini, D., Miglio, A., Salaris, M., et al. 2015, MNRAS, 453, 2290

Catelan, M. 2007, AIP Conf. Proc. Ser. 930,; College Park, MA: AIP) 39

Clayton, D. D. 1968, Principles of Stellar Evolution and Nucleosynthesis (Chicago, IL: Univ. Chicago Press)

Clayton, D. D. 1969, PhT, 22, 28

Gallart, C. 1998, ApJ, 495, L43

Gautschy, A., & Althaus, L. G. 2007, A&A, 471, 911

Girardi, L., Groenewegen, M. A. T., Hatziminaoglou, E., & da Costa, L. 2005, A&A, 436, 895

Herwig, F. 2005, ARA&A, 43, 435

Iben, I. 1975, ApJ, 196, 525

Lattanzio, J., & Karakas, A. 2016, JPhCS, 728, 022002

Marigo, P., Bressan, A., Nanni, A., Girardi, L., & Pumo, M. L. 2013, MNRAS, 434, 488

Marigo, P., Girardi, L., Bressan, A., et al. 2017, ApJ, 835, 77

Mazzitelli, I., & Dantona, F. 1986, ApJ, 308, 706

McDonald, I., De Beck, E., Zijlstra, A. A., & Lagadec, E. 2018, MNRAS, 481, 4984

Pinsonneault, M. H., Elsworth, Y., Epstein, C., et al. 2014, ApJS, 215, 19

Salaris, M., & Cassisi, S. 2006, Evolution of Stars and Stellar Populations (New York: Wiley)

Schwarzschild, M., & Härm, R. 1965, ApJ, 142, 855

Arlette Noels-Grotsch and Andrea Miglio

Chapter 8

On the Way to White Dwarf Cooling

Why does the AGB remnant become a white dwarf?

When the H-shell reaches layers very close to the surface, with a remaining envelope mass of the order of 10^{-3} $M_\odot$, it loses its ability to reignite and expand and thermal pulses do not occur anymore. The number of thermal pulses is thus strongly dependent on the mass loss.

With a negligible amount of hydrogen-rich matter on top of the H-shell, the post AGB phase starts and the evolution proceeds toward higher effective temperatures as can be seen in Figure 7.11, which shows the evolutionary track in the HR diagram of a 2 $M_\odot$ from the ZAMS to the cooling sequence of the white dwarf phase. The whole star is now contracting at a quasi constant luminosity and increasing effective temperature. During this new crossing of the HR diagram, the expanding matter ejected during the phase of super-wind can be ionized by photons arising from the star. This takes place at $T_e \sim 30,000$ K and is at the origin of the superb celestial objects known as planetary nebulae.

The very thin H-shell is still active during the whole phase of increasing effective temperature and the highest T_e marks the switching off of the H-shell. The contracting star is now devoid of any nuclear energy source and a global cooling ensues. However, in the early phase of the global contraction, He-rich layers can be heated enough to produce a final thermal runaway. The star is momentarily revitalized and becomes a *born-again* AGB star. The evolution leading to such a born-again event is shown as a blue track in Figure 7.11 shifted in L and T_e for clarity.

Our star is now facing an endless phase of cooling as a white dwarf with or without hydrogen at the surface depending of the amount of matter ejected in the post AGB phase, in particular depending on the occurrence or not of a born-again event.

doi:10.1088/2514-3433/adcf15ch8   8-1   © IOP Publishing Ltd 2025. All rights,

8.1 List of Questions

Why does the AGB remnant become a white dwarf?

Chapter 9

Epilogue

Most of the assertions made so far can and will be tested through asteroseismic analyses, which means that theory is no longer a "safe" place where you can dream without any concerns about being proven wrong. The asteroseismic issue is that we can hope to be proven right for some aspects of stellar evolution but be severely challenged for others.

doi:10.1088/2514-3433/adcf15ch9

Arlette Noels-Grotsch and Andrea Miglio

Chapter 10

Asteroseismology of Red Giant Stars

In the previous chapters, we have laid out the foundations to the inner workings of stars, highlighting open questions that pertain in particular to the red giant evolutionary phase. As stressed in Chapter 5, stars with different evolutionary states, masses, and metallicities end up sharing similar surface properties. Consequently, constraints orthogonal to photospheric temperature and luminosity are increasingly valuable for disentangling the intrinsically varied yet seemingly uniform population of red giant stars.

In this context, the detection and analysis of global resonant oscillation modes in giant stars enable us to infer both their fundamental parameters, such as mass and age, and details of their internal structure. This, in turn, provides stringent tests for many aspects of stellar physics discussed earlier.

In this chapter, we will take the following approach: first, we will succinctly introduce the main concepts necessary to understand the physical nature of pulsation modes observed in giant stars and the information they carry (Section 10.1). We will then examine how these constraints contribute to inferring global stellar properties (Section 10.2) and to testing our understanding of stellar evolution and internal transport processes (Section 10.3).

10.1 Global Oscillation Modes in Red Giant Stars

Red giant stars show rich spectra of solar-like oscillations, which are global oscillation modes excited and intrinsically damped by turbulence in the outermost layers of convective envelopes. These oscillations are in fact ubiquitous in stars with sufficiently deep convective envelopes, hence they can be detected in low- and intermediate-mass stars belonging to different evolutionary stages (main sequence, subgiant-phase, along the evolution on the red-giant branch, in the core-He burning phase and the AGB).

Moreover, solar-like oscillations exhibit remarkably rich yet relatively simple frequency patterns, which can be robustly interpreted with the aid of well-established stellar pulsation theory. The information on global and local properties of stellar interiors that can be extracted from such spectra is only briefly summarized here. For a more in-depth discussion, we encourage the keen and curious reader to consult comprehensive reviews on the subject (see Chaplin & Miglio 2013; Hekker & Christensen-Dalsgaard 2017; García & Ballot 2019).

The oscillations originate from two types of standing waves: those that are predominantly acoustic in nature —commonly referred to as pressure modes (p modes), where pressure gradients act as the restoring force —and internal gravity waves (g modes), where buoyancy plays a significant role. Additionally, mixed modes can exist, exhibiting g-mode-like behavior in the stellar core and p-mode-like behavior in the envelope. These mixed modes are typically detectable in red giants, making asteroseismic inferences particularly effective for probing near-core conditions, such as density distribution, chemical composition gradients, and internal velocity fields. Although oscillation spectra are rich and exhibit multiple overtones, geometric cancelation limits observations to modes with low angular (spherical) degree, l.

The observed mode powers in the oscillation spectrum are modulated by a Gaussian-like envelope (see top panel in Figure 10.1). The frequency of maximum oscillation power, ν_{max}, provides diagnostic insight into excitation, damping, and near-surface conditions. The wave behavior near the surface is strongly influenced by the acoustic cut-off frequency, ν_{ac}, given by

$$\nu_{\mathrm{ac}}^2 = \left(\frac{c}{4\pi H}\right)^2 \left(1 - 2\frac{dH}{dr}\right), \tag{10.1}$$

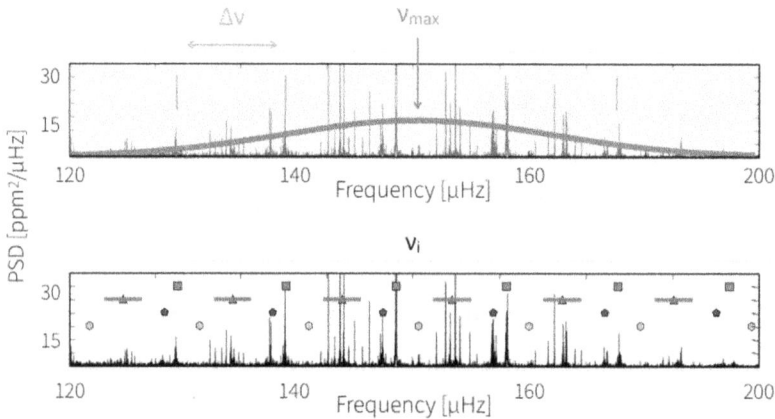

Figure 10.1. Power spectra derived from Kepler's photometric data for the red giant star KIC12008916. The upper panel presents examples of the mean seismic parameters, including the large frequency separation ($\Delta\nu$) between radial modes (see Equation (10.4)). It also highlights the oscillation envelope of power (thick red line), with its peak marked as ν_{max}. The lower panel reveals a complex and detailed spectrum of individual mode frequencies, which, despite their richness, can be reliably analysed to extract insights into both global and local stellar characteristics. Different symbols indicate various mode types: radial modes (red squares), dipole mixed pressure-gravity modes (green triangles with extended bars), quadrupole modes (blue pentagons), and octupole modes (yellow hexagons). Reproduced from Miglio et al. (2021). CC BY 4.0.

where c is the sound speed and $H = -(d, \ln\rho/dr)^{-1}$ is the density scale height. The sharp rise in ν_{ac} near the surface effectively reflects waves with $\nu < \nu_{ac}$, setting an upper frequency limit for trapped oscillations.

Brown et al. (1991) proposed that $\nu_{max} \propto \nu_{ac}$, as both are governed by near-surface properties. This leads to a scaling relation for ν_{max} in terms of measurable surface parameters, assuming an isothermal approximation of Equation (10.1), where $\nu_{ac} = c/(4\pi H)$. This results in Kjeldsen & Bedding (1995):

$$\nu_{max} \propto \nu_{ac} \propto \frac{c}{H} \propto g \; T_{eff}^{-1/2}, \tag{10.2}$$

where $g \propto M/R^2$ is the surface gravity and T_{eff} is the effective temperature. As a solar-type star evolves, its dominant oscillation frequencies decrease, primarily due to the decline in surface gravity, as shown in Figure 10.2.

Although Equation (10.2) performs well in practice, further theoretical work is needed to fully understand the observed scaling of ν_{max} (e.g., Belkacem et al. 2011).

To provide insight into the relationship between characteristic frequency patterns and the properties of their propagation cavity, i.e., the star itself, we begin by illustrating the principal features of the spectra using an asymptotic approach. Despite its limitations, the asymptotic formalism establishes fundamental concepts for interpreting resonant mode frequencies, highlighting the crucial role of astero-seismic constraints in advancing our understanding of red giant structure and evolution.

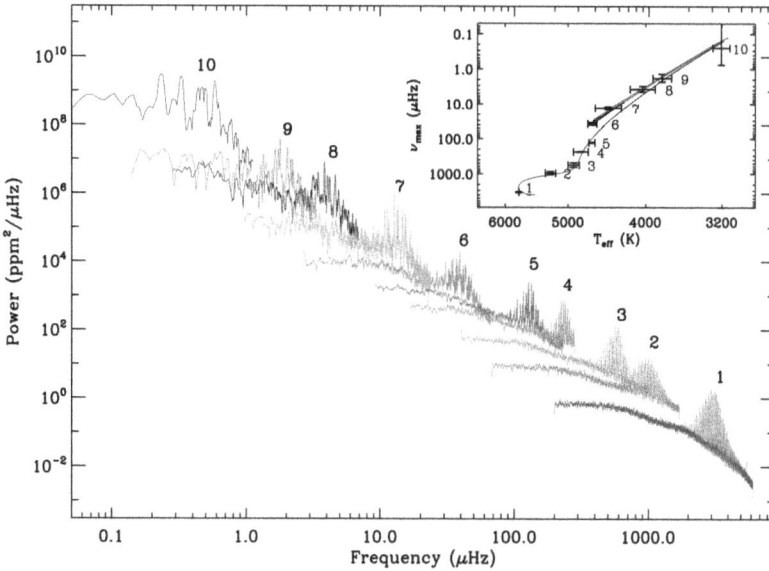

Figure 10.2. A sequence of stellar oscillation spectra arranged by decreasing ν_{max}, and consequently, decreasing surface gravity (see Equation (10.2)). The progression begins with the Sun (lower-right corner) and extends nearly up to the tip of the red giant branch (upper-left corner). Reproduced from García & Ballot (2019), with permission of the Cambridge Univerisity Press through PLSclear.

When considering p modes of high overtone numbers (radial orders) n, the asymptotic theory may be applied to describe their frequencies ν_{nl}. An approximate expression given to second order may be written (e.g., see Gough 1986):

$$\nu_{nl} \simeq \Delta\nu\left(n + \frac{l}{2} + \varepsilon\right), \tag{10.3}$$

where

$$\Delta\nu = \left(2 \int_0^R \frac{dr}{c}\right)^{-1} \tag{10.4}$$

is the inverse of the acoustic diameter, i.e., the sound travel time across a stellar diameter, with c being the sound speed and R the stellar radius, and the coefficient ε depends on the cavity boundary conditions, the behavior close to the stellar surface being most important.

Equation (10.3) thus reveals the characteristic comb-like structure of a p-mode spectrum, primarily defined by the so-called large frequency separation between modes of the same l, i.e., $\Delta\nu_{nl} = \nu_{nl} - \nu_{n-1\,l} \simeq \Delta\nu$. Additionally, modes of even and odd degrees are expected to be separated by approximately $\simeq \Delta\nu/2$.

Measuring an average large separation, $\langle\Delta\nu_{nl}\rangle$, from the oscillation spectrum provides thus a direct estimate of the star's acoustic radius. It can be shown that this quantity scales, to a very good approximation, with the square root of the mean density (Ulrich 1986), i.e.,

$$\langle\Delta\nu_{nl}\rangle \propto \langle\rho\rangle^{1/2}, \tag{10.5}$$

since the frequencies and overtone spacings are related to the dynamical timescale of the star. By combining the precise mean density estimate from $\Delta\nu$ with the information from $\nu_{\max}$ (Equation (10.2)), one can infer a star's mass and radius. This approach relies on the so-called global seismic parameters and is generally less precise and accurate than inferences based on individual mode frequencies (see Section 10.2), yet it offers a rapid and practical first estimate of a star's mass and radius—provided that model-predicted $\langle\Delta\nu_{nl}\rangle$ are used as to infer the mean density (e.g., Rodrigues et al. 2017), rather than a simplistic scaling relation.

While observable oscillations in solar-type main-sequence stars are predominantly acoustic, their spectra become more complex after hydrogen core burning ceases. This occurs because the clear separation between p-mode and g-mode frequency ranges disappears. The behavior of g modes is largely governed by the buoyancy (Brunt–Väisälä) frequency, N, given by

$$N^2 = g\left(\frac{1}{\Gamma_1}\frac{d\ln p}{dr} - \frac{d\ln\rho}{dr}\right). \tag{10.6}$$

The g modes have frequencies that are lower than N and high-order g modes may be described by an asymptotic relation in the periods, Π_{nl}, i.e.,

$$\Pi_{nl} = \nu_{nl}^{-1} \simeq \Delta\Pi_l(n + \varepsilon_g), \tag{10.7}$$

where the period separation $\Delta\Pi_l$ (analagous to $\Delta\nu$ for p modes) is given by

$$\Delta\Pi_l = \frac{2\pi^2}{\sqrt{l(l+1)}}\left(\int_{r_1}^{r_2} N\frac{dr}{r}\right)^{-1}, \tag{10.8}$$

assuming that $N^2 \geqslant 0$ in the convectively stable region bounded by $[r_1, r_2]$, with $N = 0$ at r_1 and r_2 (Tassoul 1980).

Once the central hydrogen is exhausted, the buoyancy frequency in the deep stellar interior increases significantly, extending into the frequency range of high-order p modes. When the frequency of a g mode approaches that of a non-radial p mode of the same degree, l, the two modes experience an "avoided crossing" (Aizenman et al. 1977), analogous to avoided crossings in atomic energy states. These interactions alter the mode frequencies and modify their intrinsic properties, giving rise to mixed or coupled modes—exhibiting g-mode behavior in the deep interior and p-mode behavior in the envelope.

However, not all p, g, and coupled p–g modes are necessarily observable in a spectrum. The detectability of oscillation modes can be linked to their mode inertia, which reflects the fraction of the star's mass involved in pulsation (e.g., see Aerts et al. 2010). g-dominated mixed modes, with large amplitudes in the dense core, have high inertia, making them difficult to detect. In contrast, p-dominated modes, with the largest amplitudes in the envelope, are easily observable.

Red giants have spectra of non-radial models whose patterns are dominated by the interaction between pressure and gravity modes, with the strength of such interaction (the mode coupling) strongly dependent on the density contrast between the gravity mode and pressure mode cavities, hence—to a first approximation—on the star's evolutionary state.

Moreover, while during the main sequence we observe one $l = 1$ mode per radial order, in the red-giant phase the spectrum becomes much denser with g-modes, leading to many observable $l = 1$ mixed modes per order, provided such modes have low enough inertia to be detected (see Figures 10.1 and 10.3).

In Section 10.2.4, we shall discuss the use of the g-mode period spacings as a diagnostic of the evolutionary state and see that increased period spacings in RC stars allows them to be distinguished from RGB stars that lie in close proximity in the Hertzsprung–Russell diagram. Figure 10.3 shows an example of how, in RC stars, $l = 1$ strongly coupled mixed modes with a $\Delta\Pi_1 \simeq 300$ s produce an oscillation spectrum that is significantly distinct from that of RGB stars with similar surface properties. In RGB stars, the coupling between p and g modes is weak, resulting in a simple, comb-like pattern typical of predominantly p modes in the spectrum, while the high inertias of closely spaced, g-dominated modes make them undetectable.

Up until now, we have assumed a spherically symmetric star, where oscillation frequencies are degenerate across different azimuthal orders, m. However, in the presence of rotation, this degeneracy is lifted. Rotation causes a splitting of the oscillation frequencies ν_{nl}, making the frequencies of non-radial modes ($l > 0$)

(b) KIC 5024043. "Clean dipole" star.

(c) KIC 5024327. Red Clump star.

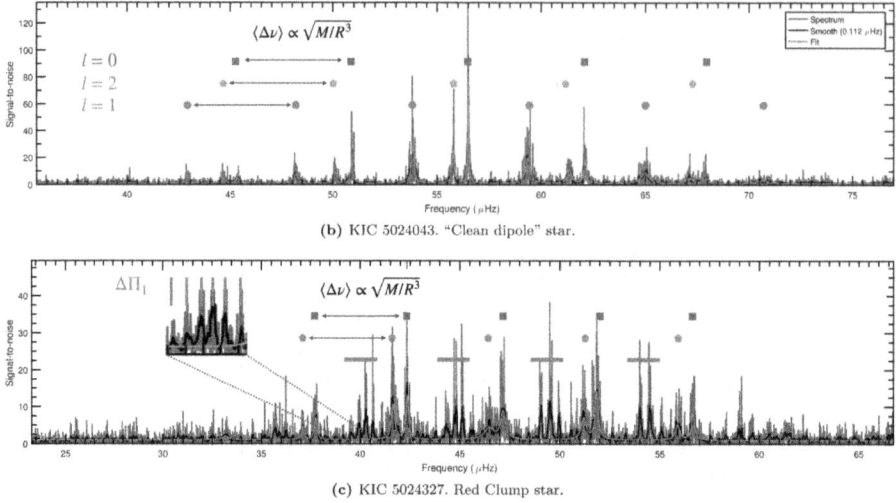

Figure 10.3. Power spectra of two stars from the same cluster, NGC 6819, with similar ν_{max}, but at different evolutionary stages—one on the red giant branch (RGB, upper panel) and the other in the red clump (RC, lower panel). The RGB star exhibits radial and predominantly acoustic non-radial modes, characterized by an approximately uniform frequency spacing ($\Delta\nu$). In contrast, the RC star displays dipolar modes with a mixed acoustic and gravity-wave nature, forming patterns associated with the period spacing of gravity modes ($\Delta\Pi_1$), which provides direct insights into the core structure of these stars. Reproduced with permission from Handberg et al. (2017).

dependent on m. For typical rotation rates in solar-like oscillators, the rotationally split frequencies, ν_{nlm}, can be written as

$$\nu_{nlm} \equiv \nu_{nl} + \delta\nu_{nlm}, \qquad (10.9)$$

with

$$\delta\nu_{nlm} \simeq \frac{m}{2\pi} \int_0^R \int_0^\pi K_{nlm}(r, \theta)\Omega(r, \theta)r\, dr\, d\theta. \qquad (10.10)$$

Here, $\Omega(r, \theta)$ is the position-dependent internal angular velocity (in radius r and co-latitude θ), and K_{nlm} is a weighting kernel that indicates how sensitive the mode is to internal rotation. For predominantly p modes, K_{nlm} is mainly localized in the stellar envelope. In contrast, for strongly coupled or g modes, the rotational splittings probe the rotation of the dense radiative core, allowing us to infer the rotational profile in the star's deepest regions (see Section 10.3). We note that the formulation presented here neglects contributions from magnetic fields, which cause asymmetries in the observed splittings (Bugnet et al. 2021; Deheuvels et al. 2022).

While the theoretical framework allows for a wide range of inferences to be made about stellar properties, it is important to remember that the precision and robustness with which one can achieve such detailed constraints on stars is also strongly dependent on the length and quality of the photometric time-series available. The analysis requires, for instance, long-duration time series to obtain the requisite frequency resolution for extracting clear signatures of rotation in the oscillation spectrum. Examples of how the different duration of a data set affects the seismic inferences is presented, e.g., in Davies & Miglio (2016) and, extensively, in Mosser et al. (2019).

10.2 Inference on *Global Stellar Properties*

When combined with measurements of a star's $T_{\rm eff}$ and photospheric chemical composition, asteroseismic data enable high-precision inferences of global stellar properties—such as mean densities, masses, radii, and ages—that are typically unattainable without seismic analysis.

Such inferences may be obtained via detailed forward and inverse modeling techniques which use different combinations of asteroseismic constraints: from average seismic parameters to individual-mode frequencies. The precision of such inferences depends on the constraints available (see Figure 10.4 for an example of the expected precision), which may vary depending on the evolutionary state and, crucially, on the quality of the data available.

As shown in Figure 10.5, observational constraints on $\Delta\nu$, $\nu_{\rm max}$, and $\Delta\Pi_1$ can, in principle, impose exceptionally tight limits on global stellar properties such as mass, radius, luminosity, and consequently distance. However, it is essential to recognize that leveraging these constraints may be hindered by limitations in current stellar models. Testing the accuracy of these percent-level inferences—particularly for mass and age, in addition to radius—remains a critical challenge, currently addressed through comparisons with independent mass determinations from stars in clusters (see Figure 10.6) and eclipsing binaries, which provide largely model-independent mass measurements (see, e.g., Gaulme et al. 2016; Brogaard et al. 2018).

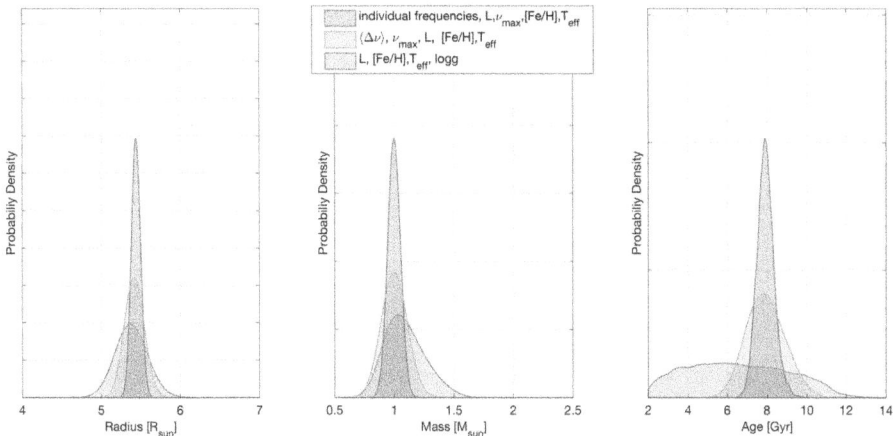

Figure 10.4. An example illustrating how the posterior probability density function for stellar radius (left panel), mass (middle panel), and age (right panel) depends on different combinations of seismic, spectroscopic, and astrometric constraints for an RGB star with $\nu_{\rm max} \simeq 110\,\mu$Hz. The analysis assumes an observation period of 150 days. Incorporating individual mode frequencies significantly enhances the precision of the inferred stellar properties. Reproduced from Miglio et al. (2021). CC BY 4.0.

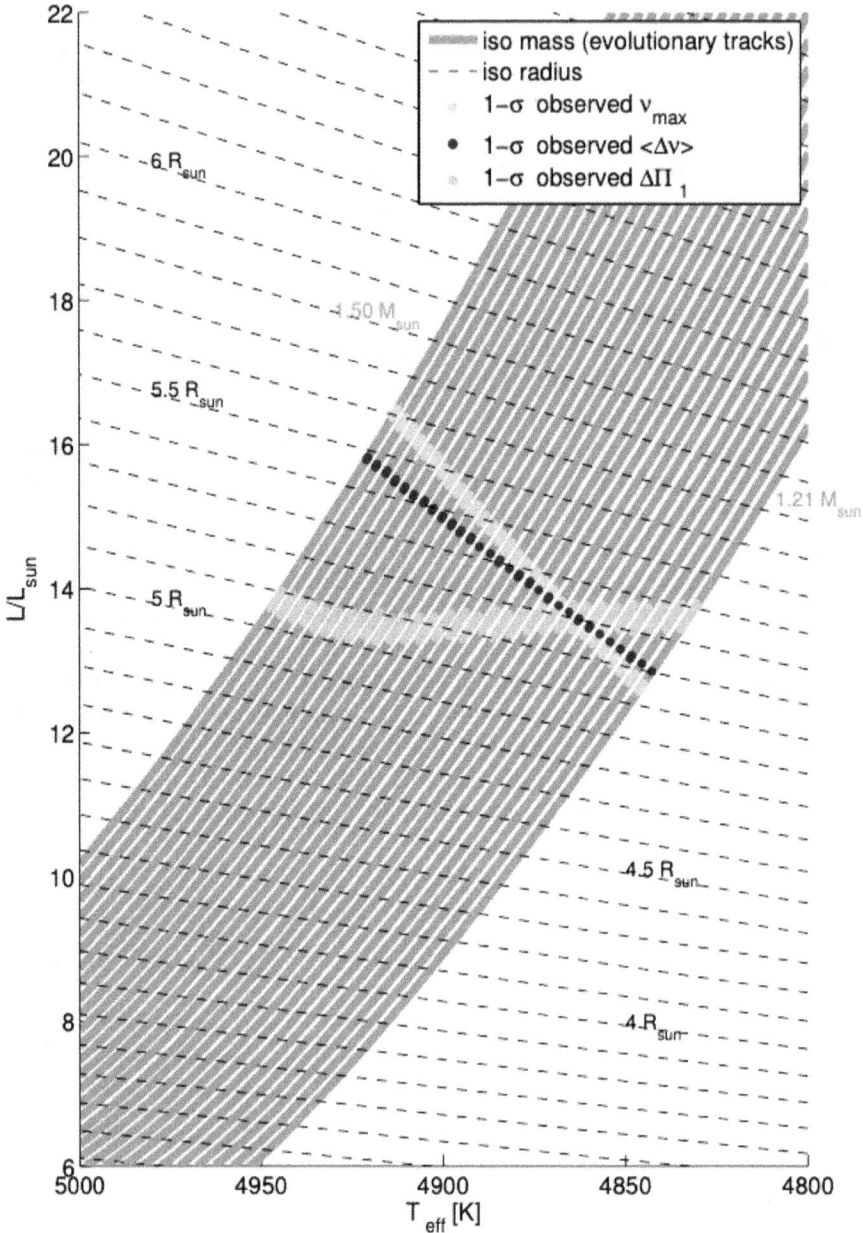

Figure 10.5. HR diagram illustrating stellar properties that satisfy the combined predictions of stellar evolutionary tracks ($M = 1.21$–1.50 $M_{\odot}$ in 0.01 $M_{\odot}$ increments, shown as solid green lines) and the asteroseismic constraints available for KIC12008916. Dashed black lines indicate contours of constant radius, spaced in steps of 0.1 $R_{\odot}$. As evident from the plot, this combination enables an exceptionally precise—though model-dependent—determination of stellar properties. Davies & Miglio (2016). John Wiley & Sons. © 2016 WILEY-VCH Verlag GmbH & Co. KGaA, Weinheim.

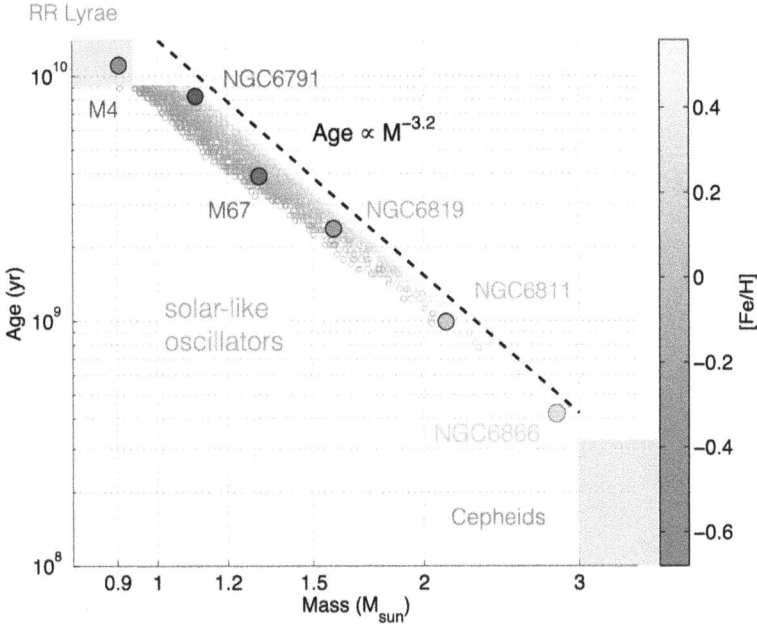

Figure 10.6. The age–mass–metallicity relation for red giants in a synthetic stellar population (Girardi et al. 2005) representing thin-disk red-giant-branch stars observed by *Kepler*. The dashed line marks the average power–law relation between age and mass for RGB stars, as also shown in Figure 5.1. Due to their broad mass range and well-defined age–mass correlations, solar-like oscillating giants (dots) serve as key tracers of the Milky Way's evolutionary history. As discussed in Section 10.2, the validation of the asteroseismic mass and age scale is anchored primarily to eclipsing binaries with solar-like oscillating components and red giants in stellar clusters. The plot specifically highlights clusters observed by the *Kepler* and K2 space missions. The diagram also highlights classical pulsators in similar evolutionary stages, such as Cepheids and RR Lyrae stars.

Here, we mention examples on how such constraints on global stellar properties can help address some of the long-standing open questions mentioned in Chapter 5, particularly when leveraging population-level constraints available for stars of different masses (and therefore ages), chemical compositions, and evolutionary stages.

To visualize this, we show in Figure 10.7 how the classical Hertzsprung–Russell diagram (as shown in Figure 4.6) can be augmented when including information about stellar masses and evolutionary state in the specific case of a sample of giants observed by *Kepler*. The large number of stars enables overdensities in the HRD to be studied, including their dependence on mass (age) and chemical composition. Moreover, as introduced in Section 10.1, the period spacing characterizing gravity modes may be used to identify core-He-burning stars among those populating the giant branches independently from photospheric constraints (see also Figure 10.11).

Such studies can help constrain internal processes, such as mixing, by examining their effect on the distribution of global stellar properties. Moreover, they can be applied across a significantly broader parameter space than what is accessible in Milky Way star clusters, as we will discuss in Section 10.2.4.

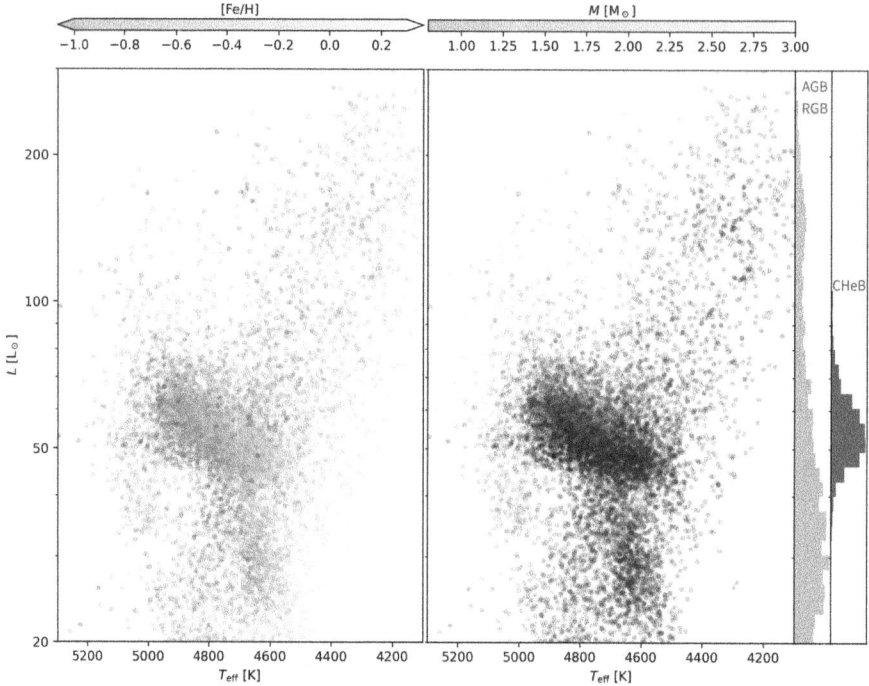

Figure 10.7. Hertzsprung–Russell diagram of red giants observed by *Kepler*. The luminosities are derived from Gaia DR3 parallaxes (Gaia Collaboration et al. 2023), while photospheric temperatures and compositions are from the APOGEE spectroscopic survey (Abdurro'uf et al. 2022). Asteroseismic constraints are taken from Yu et al. (2018). The right panel displays, using color, the mass information inferred from asteroseismology, along with the luminosity distribution of stars classified as core helium-burning, red giant branch or asymptotic giant branch. One can easily identify overdensities which are associated with various features in the evolutionary sequences (e.g., RGB bump, RC, secondary clump, AGB bump), as discussed in Chapters 5 and 6. These features and their dependence on mass and chemical composition (left panel) provide a means for detailed comparisons between the observed data and theoretical model predictions.

10.2.1 Radii, Distances, and Masses

Seismically inferred radii, combined with effective temperature, enable the determination of luminosities and, consequently, distances, with typical uncertainties of a few percent. This makes them competitive with Gaia's measurements for faint targets (Huber et al. 2017; Miglio et al. 2017). The seismic distance inference is simply based on the ratio between the star's bolometric luminosity ($L = 4\pi\sigma R^2 T_{\text{eff}}^4$), with R inferred from seismic constraints, and the apparent luminosity measured from the apparent bolometric magnitude, provided an estimate of extinction is available or inferred using multi-band photometry and T_{eff} (see Rodrigues et al. 2017). Quantitative comparisons between distances obtained with independent methods are particularly valuable as they may be used to quantify systematic effects in parallax measurements and/or in seismically inferred stellar parameters (Khan

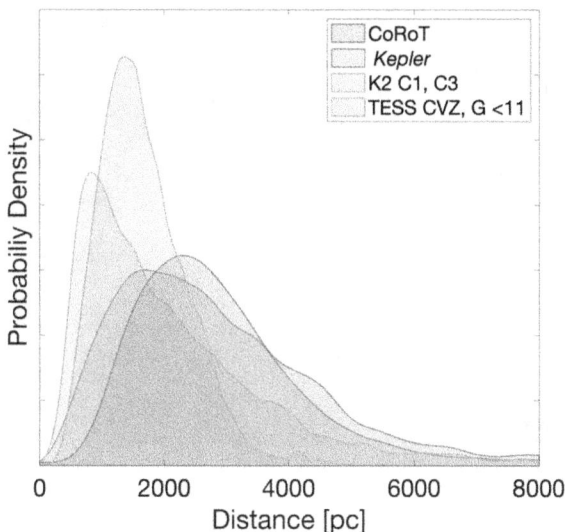

Figure 10.8. Distribution of distances for targets in various asteroseismic missions. The differing observation durations, combined with the mission-specific target selection functions, account for the variations in the distributions. Longer observations enable the detection of oscillations in longer-period stars, which are generally intrinsically brighter and more distant.

et al. 2019). The distance distribution of red giants from different asteroseismic samples is shown in Figure 10.8.

While asteroseismic constraints provide an independent, precise, and largely model-independent method for inferring stellar radii, the ability to infer masses seismically represents a truly novel advancement. Previously, masses for giant stars could only be determined in a limited number of binary systems or through comparisons with stellar model tracks and isochrones in clusters. Below, we briefly discuss how the availability of precise masses can improve age inferences, constrain processes related to mass loss, and identify likely products of binary interaction.

10.2.2 Ages

Stellar mass is a particularly valuable constraint in the case of giants, since for these stars age is primarily a function of mass. The age of low-mass red-giant stars is largely determined by the time spent on the main sequence, as thoroughly explained in Section 5.1, hence by the initial mass of the red giant's progenitor.

Our ability to infer precise (few percent) masses for tens of thousands of red giants using asteroseismology signified a fundamental step forward in the challenging task of inferring precise and accurate stellar ages (Soderblom 2010). Moreover, CoRoT, *Kepler*, K2, and TESS could detect oscillations in tens of thousands of field stars in various regions of the Milky Way (see Figure 10.9) which, combined with spectroscopic and astrometric information, are providing a reliable age scale key to inferring the formation and evolution of the Milky Way through Galactic archaeology.

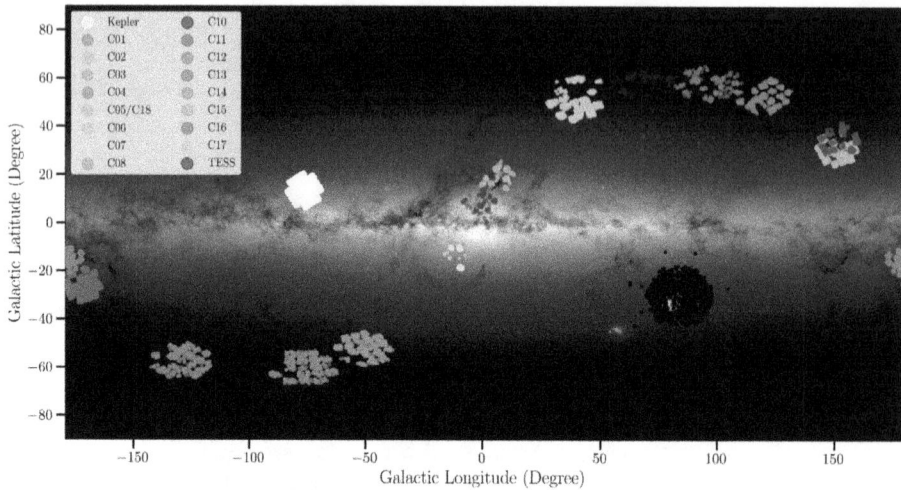

Figure 10.9. A skymap in Galactic coordinates showing the locations and coverage resulting from the crossmatch of targets observed by *Kepler*, K2, and in the TESS southern continuous viewing zone, along with the spectroscopic survey APOGEE (Abdurro'uf et al. 2022). This figure was created using the Python package mw-plot (http://milkyway-plot.readthedocs.io/), with the background image provided by ESA/Gaia/DPAC. Reproduced from Khan et al. (2023). CC BY 4.0.

Particularly important in this line of research are observational constraints linking velocity dispersion and metallicity to age in different parts of the Galaxy, as well as spatial gradients of metallicity and key abundance ratios at different ages, which reveal the signature of the Galaxy's assembly history and chemo-dynamical evolution.

Evolved low-mass stars showing solar-like oscillations, in particular, represent ideal clocks to infer the chronology of structure formation in the Milky Way, allowing precise age-dating of the oldest objects in the Galaxy (e.g., see Montalbán et al. 2021). Such precise age-dating is possible in particular when utilizing the information from individual mode frequencies, reaching the key 10% precision threshold (see Figure 10.4), providing the time resolution needed to connect meaningfully the high-redshift picture of galaxies with that given by the local Universe, in particular by stars within our Galaxy for which we have exquisitely high-resolution information about their dynamical and chemical composition.

Ages inferred using seismic constraints are now often assumed as the reference and adopted as training sets to extend the age inference to hundreds of thousands of stars using machine learning and other data-driven methods. Independent tests of the asteroseismic age scale are thus of paramount importance as they serve as the prime calibrators to what is effectively becoming the cornerstone of the stellar age scale in astrophysics, enabling us to chronologically resolve cosmic structures. The improvement in age resolution made possible by asteroseismic investigations has far-reaching implications, akin to the development of instruments that provide higher spatial or spectral resolution or extend our observational reach deeper into the cosmos. Just as these advancements have transformed our understanding of the

Universe, refining the temporal resolution enhances our ability to trace the formation and evolution of stars, galaxies, and large-scale cosmic structures with unprecedented precision.

10.2.3 Mass Loss and Gain

As mentioned earlier in this chapter and in Section 5.1, the ages of stars in the red-giant phase are determined primarily by their initial mass. However, one limitation to using measured mass as an age proxy of giant stars is the possible difference between the current and initial stellar mass (e.g., due to mass loss along the red-giant branch or by the occurrence of mass exchange and coalescence in binary systems, see, e.g., De Marco & Izzard 2017 for a review). Constraints on the efficiency of mass loss is therefore crucial to enable the accurate determination of ages of RC stars. In addition to enabling robust age estimates, setting constraints on the efficiency of mass loss on the RGB has significant implications for our understanding of the dynamical evolution of planetary systems, including our own (e.g., see Schröder & Smith 2008), as well as for understanding the physical parameters shaping the horizontal branch in globular clusters (see Section 6.2.4) and influencing evolution on the AGB (see Section 7.2).

The ability to measure the masses of red giant stars in different evolutionary states and metallicities provides a means to quantify integrated mass loss. By comparing the mass distribution of nearly coeval stars in the red clump (RC)—which have experienced mass loss in the high-luminosity, low-surface gravity portion of the RGB near the tip—with those of low-luminosity RGB stars (see, e.g., Figure 10.10), we can infer the RGB-integrated mass loss. Evidence of mass loss is now available from a few clusters and the old population in the Milky Way (Brogaard et al. 2024), offering key constraints that will inform theoretical developments on the mechanisms behind mass loss on the RGB. This work is paving the way for transforming our understanding of mass loss from a simple calibration to a well-defined physical process.

The direct measurement of stellar masses could provide a better understanding of the observational abundance of blue stragglers and other types of rejuvenated stars, i.e., stars that previously accreted material from a companion star through a stable mass transfer event or a merger.

Moreover thanks to constraints on stellar mass from asteroseismology in the open clusters observed by *Kepler*/K2 provide clear evidence for over- and under-massive stars, likely the result of mass exchange with a companion (Handberg et al. 2017; Brogaard et al. 2021). These stars end up as rejuvenated or partially stripped objects, with masses that would imply ages far exceeding the age of the Universe if they had evolved as single stars (Matteuzzi et al. 2023). There is also likely evidence of such products among field stars that are enriched in α elements, hence supposedly belonging to an old population (Martig et al. 2014; Chiappini et al. 2015).

A quantitative characterization of the occurrence and properties of these objects provide invaluable constraints to models of binary populations, e.g., event rates of binary interaction, initial-mass-ratio/period distributions (e.g., see Moe & Di

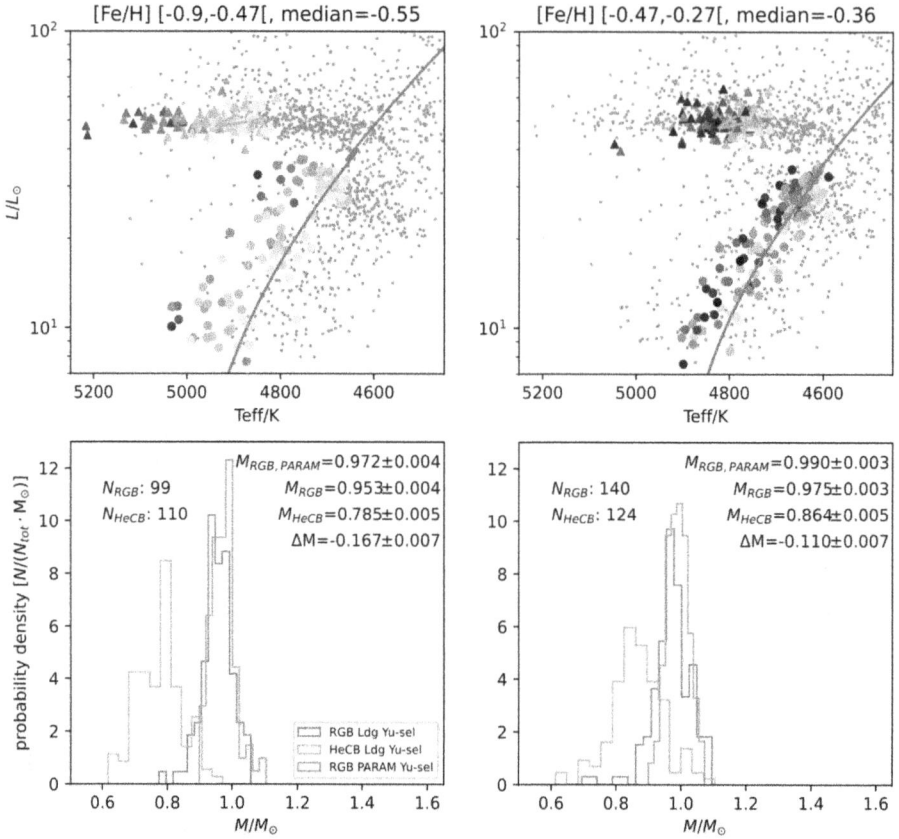

Figure 10.10. Hertzsprung–Russell diagrams and mass distributions of high-[α/Fe] stars at two different metallicities, observed by *Kepler*. The difference in median mass between low-luminosity RGB stars (circles) and RC stars (triangles) highlights the efficiency of mass loss in the luminous, low-surface-gravity region near the tip of the RGB. Reproduced from Brogaard et al. (2024). CC BY 4.0.

Stefano 2017; Izzard et al. 2018), and constraints on key parameters in the modeling of interacting binaries.

Naturally, going beyond identifying these binary products, constraints on their internal structure (see Section 10.3) offer new tools for making more direct inferences about the structural signatures that can illuminate the formation pathways of such systems.

10.2.4 Investigating Internal Structures Using HRDs Enhanced with Asteroseismic Constraints

As discussed in Chapters 5 and 6, the distribution of stars in the HRD is typically indicative of structural properties which are, however, often inaccessible to our investigation due to the degeneracy between various factors (mass, evolutionary phase, initial chemical composition). In the case of red giants, asteroseismic constraints on global stellar properties help resolve these degeneracies allowing for an indirect testing of stellar physics. For more detailed, direct tests of the internal structure, see Section 10.3.

We will here mention a couple of examples. The luminosity of the RGB bump (see Section 5.3) in principle provides tests of the maximum penetration depth of the convective envelope during the star's evolution on the RGB. The latter is strongly dependent on the efficiency of additional *internal mixing processes* happening at the boundary of the convectively unstable outer region, but also on the mass and chemical composition of the stars. A mapping of the position of the RGBb as a function of mass and metallicity (Khan et al. 2018) may thus be used to quantitatively evaluate the efficiency of such mixing, and to provide observational constraints to models of convective boundary mixing in stars (Blouin et al. 2023).

Seismic mass estimates, together with a robust inference on the evolutionary state, are also key to indirectly investigating the structure of stars in their core-He-burning phase and testing the physics that determine their distribution on the HRD (see Sections 6.1.3, 6.1.3.1, and 6.1.4). For instance, comparing the observed luminosity versus mass behavior of core-He-burning stars to those expected from models enable us to add data points to instructive diagrams, such as those in Figure 6.12, which were limited to theoretical space only until few years ago. To mention few, a detailed and accurate study of the *transition mass* between low- and intermediate-mass stars (see Section 6.1.3) opens up the possibility to constrain the helium-core mass at the He-burning ignition, thus constraining the size of the Helium core at the end of the MS (see Section 6.1.3.1). This is key to set constraints on the size of convective cores and/or additional mixing in near-core radiative regions during the MS phase, hence constraining the evolution of progenitors of red giant stars (Montalbán & Noels 2013). Second-order parameters affecting the luminosity of RC stars also become accessible to our investigation, opening up the possibility, e.g., to constrain helium and helium enrichment by comparing model-predicted and observed luminosities of RC stars of different mass and initial chemical composition (Willett 2023).

Finally, in the HRD displaying targets with seismic constraints (Figure 10.7), the AGB bump can also be identified (see Section 7.1.2). The luminosity ratio between the RC and the AGBb is crucial for constraining the size of the C–O core at the end of core-He burning, as it exhibits only weak dependence on metallicity and initial helium abundance (e.g., Bono et al. 1995). Notably, the largely unbiased sample of early AGB stars in TESS's continuous-viewing zone (Mackereth et al. 2021), combined with exquisite parallaxes from Gaia, provides key observational evidence to indirectly constrain mixing during the core-helium burning phase (see Section 7.1.3), while also enabling detailed seismic inference on the structure of such stars (see Section 7.2).

Recasting decades-long open questions in stellar physics in an environment enriched by seismic constraints is at its infancy only and exciting new finding are expected in the years to come, especially when coupled with advances in our ability to understand the physics of transport processes in stars with the aid of numerical simulations.

10.3 Direct Constraints on the *Internal Structure*

In the previous section, we provided examples of how seismic constraints can be used to infer global stellar properties—such as mass, radius, age, and evolutionary state—thereby offering key, novel insights to address unresolved questions in stellar physics. Here, we mention seismic diagnostics that can reveal localized, specific features of the internal stellar structure.

10.3.1 Mixed-Mode Patterns: Average Period Spacing and Coupling between p- and g-Cavities

To a first approximation, from the (mixed) oscillation modes patterns observed in red giants, one can infer average quantities characteristics of high-order p and g modes, i.e., respectively, large frequency spacing ($\Delta\nu$) and period spacing ($\Delta\Pi_1$). As introduced in Section 10.2, $\Delta\nu$ (p modes) is primarily a proxy for the stellar mean density, while $\Delta\Pi_1$ (see Equation (10.8)) is determined by the thermal and chemical composition stratification of the radiative, near-core regions.

First of all, $\Delta\Pi_1$ provides a direct and clear way to identify the evolutionary state of low-mass giants (see Figure 10.11), primarily due to the different density

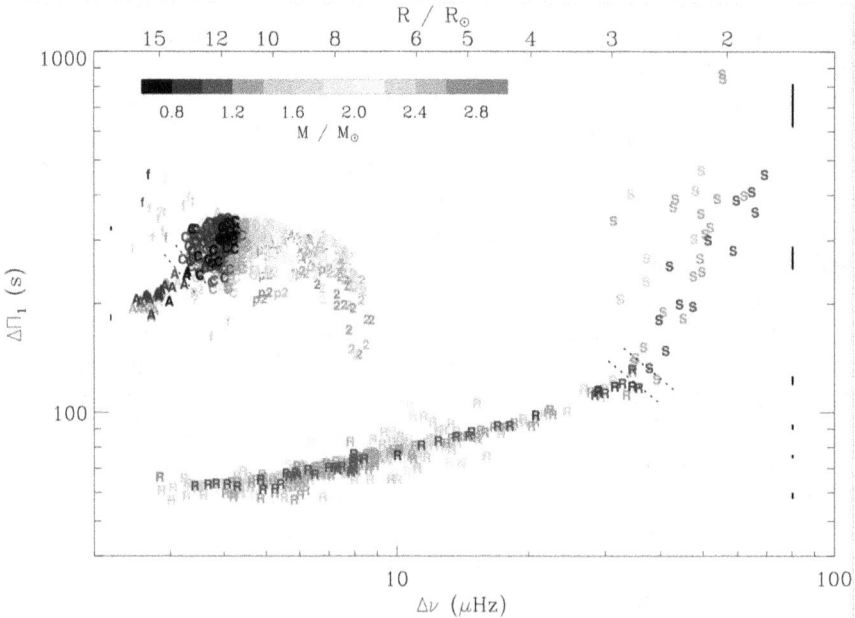

Figure 10.11. Period spacing $\Delta\Pi_1$ as a function of the acoustic-mode frequency spacing $\Delta\nu$, measured in approximately 1200 evolved stars observed by *Kepler*. The period spacing provides a direct probe of the density stratification of the stellar core, serving as a clear proxy for the evolutionary state of stars. The evolutionary states are labeled as follows: S (subgiants), R (RGB), C (red clump), 2 (secondary clump), and A (stars leaving the red clump and moving toward the AGB). The symbol "f" marks stars that are candidates for being in the inter-flash stage. The estimated seismic mass is indicated by the color code. Credit: Mosser et al, (2014), reproduced with permission © ESO.

distributions in the cores of RGB and RC stars, as well as the presence of a convective core in the latter (Bedding et al. 2011; Montalbán et al. 2010).

Moreover, precise knowledge of the average $\Delta\Pi_1$ provides, as an integrated quantity, serves as a direct probe of the near-core regions of giants, yielding information about the mass of the inert helium core on the RGB and the density and chemical composition profiles of the inner regions in core-He burning stars (e.g., Montalbán & Noels 2013; Bossini et al. 2015; Constantino et al. 2015; Cunha et al. 2015). Such constraints are particularly useful for core-He-burning stars, where uncertainties regarding the structure and evolution of the convective core and the surrounding layers (overshooting and semiconvective regions see Chapter 6) remain substantial.

Comparisons between observed $\Delta\Pi_1$ and model predictions are already helping to discriminate among the various mixing prescriptions described in Section 6.2 (see, e.g., Figure 10.12) and can be further complemented by seismic signatures of local sharp-structure variations (see Section 10.3.3).

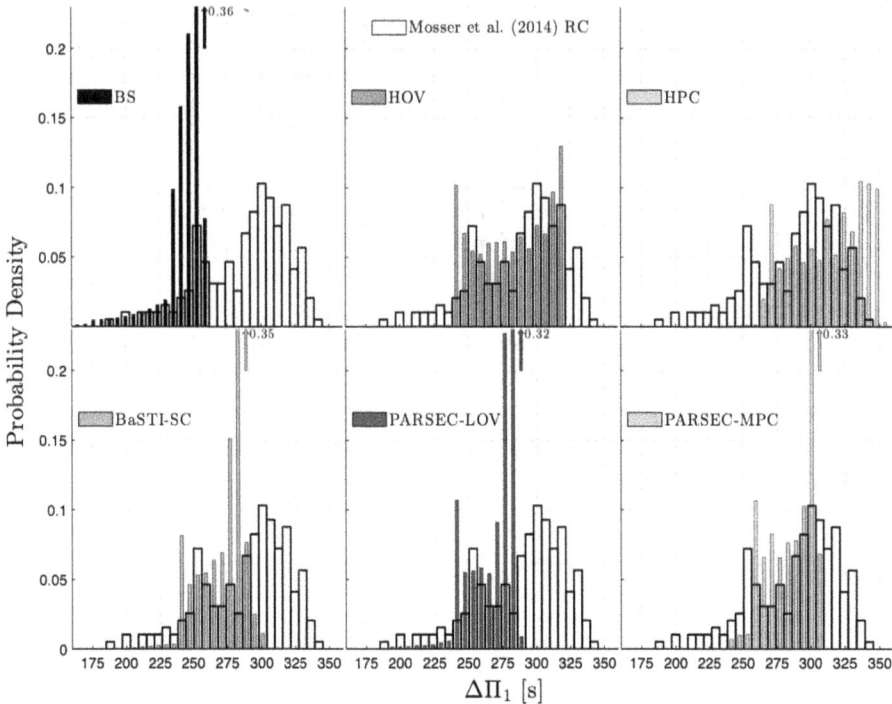

Figure 10.12. Comparison between the observed period-spacing distribution of RC stars with $M \sim 1.5\,M_\odot$ from Mosser et al. (2014) and the predictions from models assuming different prescriptions for the evolution of the convective core (see Section 6.2). While the analysis of seismic glitches can reveal details about the local chemical composition near the edge of the convective core (see Section 10.3.3), an average $\Delta\Pi_1$ already provides strong observational evidence to rule out some of the prescriptions discussed in the literature. Reproduced with permission from Bossini et al. (2015).

In addition to $\Delta\Pi_1$ the observational signature of the coupling between the pressure- and gravity-mode cavities allows one to infer properties of the evanescent region between the two coupled propagation cavities (Takata 2016; Mosser et al. 2017; Pinçon et al. 2020; van Rossem et al. 2024), offering yet another promising diagnostic for future investigations.

Finally, seismic constraints on core and envelope properties based on the mixed modes patterns are opening the door to the identification and characterization of non-standard structures, such as those that may result from binary interactions (Rui & Fuller 2021; Deheuvels et al. 2022; Matteuzzi et al. 2023).

10.3.2 Constraints on the *Internal Rotation Rate*

Rotational frequency splittings (see Section 10.1) provide critical insights into the internal rotation rate of giants. By detecting rotational splittings associated with mixed p–g modes, we can primarily measure core rotation and, with more detailed and localized inferences, probe the rotational profile of the star (Beck et al. 2012; Deheuvels et al. 2012; Mosser et al. 2012; Di Mauro et al. 2016; Gehan et al. 2018). These splittings appear in the power spectrum as dense clusters of $l = 1$ modes modes, with the mode at the center of each cluster typically dominated by p-mode characteristics, making it somewhat more sensitive to the star's outer layers. In contrast, adjacent modes exhibit more g-mode-like behavior, with rotational kernels (Equation (10.10)) sensitive to the star's central regions only (Beck et al. 2012; Deheuvels et al. 2012).

Thanks to *Kepler* data, rotational frequency splittings have been estimated for thousands of giants across a range of masses, metallicities, and evolutionary stages. This large data set has provided stringent constraints on the evolution of mean core rotation rates in stars (Gehan et al. 2018), revealing significant insights into the redistribution of internal angular momentum during the post-main-sequence phase. These results highlight the limitations of current stellar models in describing angular momentum transport, prompting the investigation of alternative mechanisms and instabilities to explain the missing transport processes (see reviews by Aerts et al. 2019, Eggenberger 2024).

10.3.3 Deviations from the Expected, Approximated, Frequency Patterns of p and g Modes

Deviations from the asymptotic mixed-mode patterns described in Section 10.1 may be used to detect and infer detailed properties of *localized gradients* in, e.g., the chemical composition/thermal stratification/sound speed profile (referred to as "glitches").

Examples of such localized features include regions of strong thermal and chemical gradients and, e.g., the He-ionization region where the adiabatic index, hence sound speed, has a local minimum, and whose signature may be used to infer the envelope He abundance using pressure modes (e.g., Vorontsov 1988; Monteiro et al. 1994; Miglio et al. 2010; Broomhall et al. 2014; McKeever et al. 2019).

Gravity modes are highly sensitive to sharp structural variations as well (Cunha et al. 2015), particularly in the near-center radiative regions. As such, the determination of glitch parameters offers deep insight into the chemical and thermal stratification of the region of mixing beyond the convective core in core-He burning stars. While observational evidence for these glitches is well established (Mosser et al. 2015; Vrard et al. 2022), their application remains in its early stages. Nevertheless, this approach shows significant promise for providing precise, localized stratigraphy that could illuminate the transport processes at the interface between the growing convective core and the surrounding radiative region (see Section 6.2).

10.4 A Bright Future for Asteroseismology and Stellar Physics

Advances in asteroseismology are revolutionizing our ability to test and refine models of stellar structure. For the first time, we have empirical tools to probe the interiors of red giants, offering insights that surface observations alone cannot provide. These constraints are not only confirming theoretical predictions but are also exposing critical limitations in current models, marking a turning point in our understanding of stellar physics.

This chapter has highlighted a few established applications of asteroseismic data, but the full potential of this field is only beginning to emerge. Future efforts promise breakthroughs also in areas such as mapping the morphology of magnetic fields in giants (Li et al. 2022; Hatt et al. 2024), studying stellar activity and the products of binary evolution (Gaulme et al. 2020; Rui & Fuller 2021), and probing the limits of fundamental physics through stars (Raffelt 1996). These advancements will challenge our understanding of key physical processes and refine our models with unprecedented precision.

Stellar physics has traditionally been viewed as a well-established discipline, capable of accurately describing the surface properties of stars. However, asteroseismology now enables us to investigate their interiors with remarkable detail, transforming stars into natural laboratories for testing physical processes at the percent level. By identifying where our models fall short, we open pathways for new discoveries and theoretical advancements.

In this exciting new era, the ability to prove our models wrong is not a limitation but a strength. It is through these challenges that stellar physics will continue to evolve, bringing us closer to a comprehensive understanding of the complex processes that govern stars and their evolution.

References

Abdurro'uf, , Accetta, K., Aerts, C., et al. 2022, ApJS, 259, 35

Aerts, C., Christensen-Dalsgaard, J., & Kurtz, D. W. 2010, Asteroseismology (Berlin: Springer Science & Business Media)

Aerts, C., Mathis, S., & Rogers, T. M. 2019, ARA&A, 57, 35

Aizenman, M., Smeyers, P., & Weigert, A. 1977, A&A, 58, 41

Beck, P. G., Montalban, J., Kallinger, T., et al. 2012, Natur, 481, 55

Bedding, T. R., Mosser, B., Huber, D., et al. 2011, Natur, 471, 608

Belkacem, K., Goupil, M. J., Dupret, M. A., et al. 2011, A&A, 530, A142

Blouin, S., Mao, H., Herwig, F., et al. 2023, MNRAS, 522, 1706

Bono, G., Castellani, V., deglaInnocenti, S., & Pulone, L. 1995, A&A, 297, 115

Bossini, D., Miglio, A., Salaris, M., et al. 2015, MNRAS, 453, 2290

Brogaard, K., Arentoft, T., Jessen-Hansen, J., & Miglio, A. 2021, MNRAS, 507, 496

Brogaard, K., Hansen, C. J., Miglio, A., et al. 2018, MNRAS, 476, 3729

Brogaard, K., Miglio, A., van Rossem, W. E., Willett, E., & Thomsen, J. S. 2024, A&A, 691, A288

Broomhall, A. M., Miglio, A., Montalbán, J., et al. 2014, MNRAS, 440, 1828

Brown, T. M., Gilliland, R. L., Noyes, R. W., & Ramsey, L. W. 1991, ApJ, 368, 599

Bugnet, L., Prat, V., Mathis, S., et al. 2021, A&A, 650, A53

Chaplin, W. J., & Miglio, A. 2013, ARA&A, 51, 353

Chiappini, C., Anders, F., Rodrigues, T. S., et al. 2015, A&A, 576, L12

Constantino, T., Campbell, S. W., Christensen-Dalsgaard, J., Lattanzio, J. C., & Stello, D. 2015, MNRAS, 452, 123

Cunha, M. S., Stello, D., Avelino, P. P., Christensen-Dalsgaard, J., & Townsend, R. H. D. 2015, ApJ, 805, 127

Davies, G. R., & Miglio, A. 2016, AN, 337, 774

De Marco, O., & Izzard, R. G. 2017, PASA, 34, e001

Deheuvels, S., Ballot, J., Gehan, C., & Mosser, B. 2022, A&A, 659, A106

Deheuvels, S., García, R. A., Chaplin, W. J., et al. 2012, ApJ, 756, 19

Di Mauro, M. P., Ventura, R., Cardini, D., et al. 2016, ApJ, 817, 65

Eggenberger, P. 2024, arXiv e-prints, arXiv:2409.11354

Gaia CollaborationVallenari, A., Brown, A. G. A., et al. 2023, A&A, 674, A1

García, R. A., & Ballot, J. 2019, LRSP, 16, 4

Gaulme, P., McKeever, J., Jackiewicz, J., et al. 2016, ApJ, 832, 121

Gaulme, P., Jackiewicz, J., Spada, F., et al. 2020, A&A, 639, A63

Gehan, C., Mosser, B., Michel, E., Samadi, R., & Kallinger, T. 2018, A&A, 616, A24

Girardi, L., Groenewegen, M. A. T., Hatziminaoglou, E., & da Costa, L. 2005, A&A, 436, 895

Gough, D. O. 1986, Hydrodynamic and Magnetodynamic Problems in the Sun and Stars, ed. Y. Osaki (Tokyo: Tokyo University) 117

Handberg, R., Brogaard, K., Miglio, A., et al. 2017, MNRAS, 472, 979

Hatt, E. J., Joel Ong, J. M., Nielsen, M. B., et al. 2024, MNRAS, 534, 1060

Hekker, S., & Christensen-Dalsgaard, J. 2017, A&AR, 25, 1

Huber, D., Zinn, J., Bojsen-Hansen, M., et al. 2017, ApJ, 844, 102

Izzard, R. G., Preece, H., Jofre, P., et al. 2018, MNRAS, 473, 2984

Khan, S., Hall, O. J., Miglio, A., et al. 2018, ApJ, 859, 156

Khan, S., Miglio, A., Mosser, B., et al. 2019, A&A, 628, A35

Khan, S., Miglio, A., Willett, E., et al. 2023, A&A, 677, A21

Kjeldsen, H., & Bedding, T. R. 1995, A&A, 293, 87

Li, G., Deheuvels, S., Ballot, J., & Lignières, F. 2022, Natur, 610, 43

Mackereth, J., Miglio, A., Elsworth, Y., et al. 2021, MNRAS, 502, 1947

Martig, M., Minchev, I., & Flynn, C. 2014, MNRAS, 443, 2452

Matteuzzi, M., Montalbán, J., Miglio, A., et al. 2023, A&A, 671, A53

McKeever, J. M., Basu, S., & Corsaro, E. 2019, ApJ, 874, 180

Miglio, A., Chiappini, C., Mosser, B., et al. 2017, AN, 338, 644

Miglio, A., Montalbán, J., Carrier, F., et al. 2010, A&A, 520, L6

Miglio, A., Girardi, L., Grundahl, F., et al. 2021, ExA, 51, 963

Moe, M., & Di Stefano, R. 2017, ApJS, 230, 15

Montalbán, J., Miglio, A., Noels, A., Scuflaire, R., & Ventura, P. 2010, ApJ, 721, L182

Montalbán, J., & Noels, A. 2013, EPJ Web Conf., 43, 03002

Montalbán, J., Ted Mackereth, J., Miglio, A., et al. 2021, NatAs, 5, 640

Monteiro, M. J. P. F. G., Christensen-Dalsgaard, J., & Thompson, M. J. 1994, A&A, 283, 247

Mosser, B., Benomar, O., Belkacem, K., et al. 2014, A&A, 572, L5

Mosser, B., Goupil, M. J., Belkacem, K., et al. 2012, A&A, 548, A10

Mosser, B., Michel, E., Samadi, R., et al. 2019, A&A, 622, A76

Mosser, B., Pinçon, C., Belkacem, K., Takata, M., & Vrard, M. 2017, A&A, 600, A1

Mosser, B., Vrard, M., Belkacem, K., Deheuvels, S., & Goupil, M. J. 2015, A&A, 584, A50

Pinçon, C., Goupil, M. J., & Belkacem, K. 2020, A&A, 634, A68

Raffelt, G. G. 1996, Stars as Laboratories for Fundamental Physics: The Astrophysics of Neutrinos, Axions, and Other Weakly Interacting Particles (Chicago, IL: Univ. Chicago Press)

Rodrigues, T. S., Bossini, D., Miglio, A., et al. 2017, MNRAS, 467, 1433

Rui, N. Z., & Fuller, J. 2021, MNRAS, 508, 1618

Schröder, K. P., & Smith, R. C. 2008, MNRAS, 386, 155

Soderblom, D. R. 2010, ARA&A, 48, 581

Takata, M. 2016, PASJ, 68, 109

Tassoul, M. 1980, ApJS, 43, 469

Ulrich, R. K. 1986, ApJ, 306, L37

van Rossem, W. E., Miglio, A., & Montalbán, J. 2024, A&A, 691, A177

Vorontsov, S. V. 1988, Advances in Helio- and Asteroseismology, IAU Symp. 123 ed. J. Christensen-Dalsgaard, & S. Frandsen (Dordrecht: Reidel Publishing) 151

Vrard, M., Cunha, M. S., Bossini, D., et al. 2022, NatCo, 13, 7553

Willett, E. 2023, PhD thesis, Univ. Birmingham Available at https://etheses.bham.ac.uk//id/eprint/14396/

Yu, J., Huber, D., Bedding, T. R., et al. 2018, ApJS, 236, 42

Bibliography

Abdurro'uf,, Accetta, K., Aerts, C., et al. 2022, ApJS, 259, 35

Adelberger, E. G., García, A., Robertson, R. G. H., et al. 2011, RvMP, 83, 195

Aerts, C., Christensen-Dalsgaar, J., & Kurtz, D. W. 2010, Asteroseismology (Berlin: Springer Science & Business Media)

Aerts, C., Mathis, S., & Rogers, T. M. 2019, ARA&A, 57, 35

Aizenman, M., Smeyers, P., & Weigert, A. 1977, A&A, 58, 41

Alongi, M., Bertelli, G., Bressan, A., & Chiosi, C. 1991, A&A, 244, 95

Angulo, C., Arnould, M., Rayet, M., et al. 1999, NuPhA, 656, 3

Arp, H. C., Baum, W. A., & Sandage, A. R. 1952, AJ, 57, 4

Asplund, M., Grevesse, N., Sauval, A. J., & Scott, P. 2009, ARA&A, 47, 481

Baglin, A., Auvergne, M., Barge, P.COROT Team, et al. 2006, The CoRoT Mission Pre-Launch Status—Stellar Seismology and Planet Finding, Vol. 1306, ed. M. Fridlund, A. Baglin, J. Lochard, & L. Conroy (Paris: European Space Agency) 33

Basu, S., & Antia, H. M. 1994, MNRAS, 269, 1137

Beck, P. G., Montalbán, J., Kallinger, T., et al. 2012, Natur, 481, 55

Bedding, T. R., Mosser, B., Huber, D., et al. 2011, Natur, 471, 608

Belkacem, K., Goupil, M. J., Dupret, M. A., et al. 2011, A&A, 530, A142

Blouin, S., Mao, H., Herwig, F., et al. 2023, MNRAS, 522, 1706

Bono, G., Castellani, V., degl'Innocenti, S., & Pulone, L. 1995, A&A, 297, 115

Bossini, D., Miglio, A., Salaris, M., et al. 2015, MNRAS, 453, 2290

Bressan, A., Marigo, P., Girardi, L., et al. 2012, MNRAS, 427, 127

Brogaard, K., Arentoft, T., Jessen-Hansen, J., & Miglio, A. 2021, MNRAS, 507, 496

Brogaard, K., Hansen, C. J., Miglio, A., et al. 2018, MNRAS, 476, 3729

Brogaard, K., Miglio, A., van Rossem, W. E., Willett, E., & Thomsen, J. S. 2024, A&A, 691, A288

Broomhall, A. M., Miglio, A., Montalbán, J., et al. 2014, MNRAS, 440, 1828

Brown, T. M., Gilliland, R. L., Noyes, R. W., & Ramsey, L. W. 1991, ApJ, 368, 599

Bugnet, L., Prat, V., Mathis, S., et al. 2021, A&A, 650, A53

Castellani, V., Degl'Innocenti, S., Girardi, L., et al. 2000, A&A, 354, 150

Castellani, V., Giannone, P., & Renzini, A. 1971, Ap&SS, 10, 355

Catelan, M. 2007, AIP Conf. Proc. Ser. (College Park, MA: AIP) 930, 39

Chaplin, W. J., & Miglio, A. 2013, ARA&A, 51, 353

Chiappini, C., Anders, F., Rodrigues, T. S., et al. 2015, A&A, 576, L12

Christensen-Dalsgaard, J. 2015, MNRAS, 453, 666

Claret, A., & Torres, G. 2018, ApJ, 859, 100

Claret, A., & Torres, G. 2019, ApJ, 876, 134

Clayton, D. D. 1968, Principles of Stellar Evolution and Nucleosynthesis (Chicago, IL: Univ. Chicago Press)

Clayton, D. D. 1969, PhT, 22, 28

Constantino, T., & Baraffe, I. 2018, A&A, 618, A177

Constantino, T., Campbell, S. W., Christensen-Dalsgaard, J., Lattanzio, J. C., & Stello, D. 2015, MNRAS, 452, 123

Cox, J. P., & Salpeter, E. E. 1964, ApJ, 140, 485

Cristini, A., Hirschi, R., Meakin, C., et al. 2019, MNRAS, 484, 4645

Cunha, M. S., Stello, D., Avelino, P. P., Christensen-Dalsgaard, J., & Townsend, R. H. D. 2015, ApJ, 805, 127

Davies, G. R., & Miglio, A. 2016, AN, 337, 774

De Marco, O., & Izzard, R. G. 2017, PASA, 34, e001

Deal, M., Richard, O., & Vauclair, S. 2016, A&A, 589, A140

Deheuvels, S., Ballot, J., Gehan, C., & Mosser, B. 2022, A&A, 659, A106

Deheuvels, S., Brandão, I., Silva Aguirre, V., et al. 2016, A&A, 589, A93

Deheuvels, S., García, R. A., Chaplin, W. J., et al. 2012, ApJ, 756, 19

Deinzer, W., & Salpeter, E. E. 1965, ApJ, 142, 813

Di Mauro, M. P., Ventura, R., Cardini, D., et al. 2016, ApJ, 817, 65

Dziembowski, W. 1977, AcA, 27, 95

Eggenberger, P. 2024, arXiv e-prints, arXiv:2409.11354

Eggenberger, P., Meynet, G., Maeder, A., et al. 2008, Ap&SS, 316, 43

Eggenberger, P., Montalbán, J., & Miglio, A. 2012, A&A, 544, L4

Faulkner, D. J., & Cannon, R. D. 1973, ApJ, 180, 435

Freytag, B., Ludwig, H. G., & Steffen, M. 1996, A&A, 313, 497

Fu, X., Bressan, A., Marigo, P., et al. 2018, MNRAS, 476, 496

Gabriel, M., & Belkacem, K. 2018, A&A, 612, A21

Gabriel, M., Noels, A., Montalbán, J., & Miglio, A. 2014, A&A, 569, A63

Gaia Collaboration, Brown, A. G. A., Vallenari, A., et al. 2016, A&A, 595, A2

Gaia Collaboration, Vallenari, A., Brown, A. G. A., et al. 2023, A&A, 674, A1

Gallart, C. 1998, ApJ, 495, L43

García, R. A., & Ballot, J. 2019, LRSP, 16, 4

Gaulme, P., McKeever, J., Jackiewicz, J., et al. 2016, ApJ, 832, 121

Gaulme, P., Jackiewicz, J., Spada, F., et al. 2020, A&A, 639, A63

Gautschy, A., & Althaus, L. G. 2007, A&A, 471, 911

Gehan, C., Mosser, B., Michel, E., Samadi, R., & Kallinger, T. 2018, A&A, 616, A24

Girardi, L. 1999, MNRAS, 308, 818

Girardi, L., Groenewegen, M. A. T., Hatziminaoglou, E., & da Costa, L. 2005, A&A, 436, 895

Gough, D. O. 1986, Hydrodynamic and Magnetodynamic Problems in the Sun and Stars, ed. Y. Osaki (Tokyo: Tokyo University) 117

Handberg, R., Brogaard, K., Miglio, A., et al. 2017, MNRAS, 472, 979

Hatt, E. J., Joel Ong, J. M., Nielsen, M. B., et al. 2024, MNRAS, 534, 1060

Hayashi, C. 1961, PASJ, 13, 450

Hayashi, C., & Nakano, T. 1963, PThPh, 30, 460

Hekker, S., & Christensen-Dalsgaard, J. 2017, A&AR, 25, 1

Herwig, F. 2005, ARA&A, 43, 435

Herwig, F., Bloecker, T., Schoenberner, D., & El Eid, M. 1997, A&A, 324, L81

Huber, D., Zinn, J., Bojsen-Hansen, M., et al. 2017, ApJ, 844, 102

Iben, I. 1975, ApJ, 196, 525

Iben, I. 2013, Stellar Evolution Physics, Volume 1: Physical Processes in Stellar Interiors (Cambridge: Cambridge Univ. Press)

Iben, I. 2013, Stellar Evolution Physics, Volume 2: Advanced Evolution of Single Stars (Cambridge: Cambridge Univ. Press)

Iglesias, C. A., & Rogers, F. J. 1996, ApJ, 464, 943

Irwin, A. W. 2012, Astrophysics Source Code Library, record ascl:1211.002

Izzard, R. G., Preece, H., Jofre, P., et al. 2018, MNRAS, 473, 2984

Khan, S., Hall, O. J., Miglio, A., et al. 2018, ApJ, 859, 156

Khan, S., Miglio, A., Mosser, B., et al. 2019, A&A, 628, A35

Khan, S., Miglio, A., Willett, E., et al. 2023, A&A, 677, A21

Kippenhahn, R., & Weigert, A. 1994, Stellar Structure and Evolution (Berlin: Springer)

Kjeldsen, H., & Bedding, T. R. 1995, A&A, 293, 87

Lagioia, E. P., Milone, A. P., Marino, A. F., et al. 2018, MNRAS, 475, 4088

Lattanzio, J., & Karakas, A. 2016, JPhCS, 728, 022002

Lattanzio, J. C., & Wood, P. R. 2004, ed. H. J. Habing, & H. Olofsson Asymptotic Giant Branch Stars. Astronomy and Astrophysics Library (New York: Springer) 23

Ledoux, P. 1947, ApJ, 105, 305

Li, G., Deheuvels, S., Ballot, J., & Lignières, F. 2022, Natur, 610, 43

Lynden-Bell, D., & Wood, R. 1968, MNRAS, 138, 495

Mackereth, J., Miglio, A., Elsworth, Y., et al. 2021, MNRAS, 502, 1947

Maeder, A. 2009, Physics, Formation and Evolution of Rotating Stars (Berlin: Springer)

Marigo, P., Bressan, A., Nanni, A., Girardi, L., & Pumo, M. L. 2013, MNRAS, 434, 488

Marigo, P., Girardi, L., Bressan, A., et al. 2017, ApJ, 835, 77

Martig, M., Minchev, I., & Flynn, C. 2014, MNRAS, 443, 2452

Matteuzzi, M., Montalbán, J., Miglio, A., et al. 2023, A&A, 671, A53

Mazzitelli, I., & Dantona, F. 1986, ApJ, 308, 706

McDonald, I., De Beck, E., Zijlstra, A. A., & Lagadec, E. 2018, MNRAS, 481, 4984

McKeever, J. M., Basu, S., & Corsaro, E. 2019, ApJ, 874, 180

Meakin, C. A., & Arnett, D. 2007, ApJ, 667, 448

Meakin, C. A., & Arnett, W. D. 2010, Ap&SS, 328, 221

Miglio, A. 2012, ApSSP, 26, 11

Miglio, A., Chiappini, C., Mosser, B., et al. 2017, AN, 338, 644

Miglio, A., Montalbán, J., Carrier, F., et al. 2010, A&A, 520, L6

Miglio, A., Montalbán, J., Noels, A., & Eggenberger, P. 2008, MNRAS, 386, 1487

Miglio, A., Girardi, L., Grundahl, F., et al. 2021, ExA, 51, 963

Moe, M., & Di Stefano, R. 2017, ApJS, 230, 15

Monelli, M., Cassisi, S., Bernard, E. J., et al. 2010, ApJ, 718, 707

Montalbán, J., Miglio, A., Noels, A., Scuflaire, R., & Ventura, P. 2010, ApJ, 721, L182

Montalbán, J., & Noels, A. 2013, EPJ Web Conf., 43, 03002

Montalbán, J., Ted Mackereth, J., Miglio, A., et al. 2021, NatAs, 5, 640

Monteiro, M. J. P. F. G., Christensen-Dalsgaard, J., & Thompson, M. J. 1994, A&A, 283, 247

Mosser, B., Benomar, O., Belkacem, K., et al. 2014, A&A, 572, L5

Mosser, B., Goupil, M. J., Belkacem, K., et al. 2012, A&A, 548, A10

Mosser, B., Michel, E., Samadi, R., et al. 2019, A&A, 622, A76

Mosser, B., Pinçon, C., Belkacem, K., Takata, M., & Vrard, M. 2017, A&A, 600, A1

Mosser, B., Vrard, M., Belkacem, K., Deheuvels, S., & Goupil, M. J. 2015, A&A, 584, A50

Nakano, T. 2014, 50 Years of Brown Dwarfs, ed. V. Joergens (Switzerland: Springer International) 5

Noels, A., Montalbán, J., Miglio, A., Godart, M., & Ventura, P. 2010, Ap&SS, 328, 227

Pinçon, C., Goupil, M. J., & Belkacem, K. 2020, A&A, 634, A68

Pinsonneault, M. H., Elsworth, Y., Epstein, C., et al. 2014, ApJS, 215, 19

Raffelt, G. G. 1996, Stars as Laboratories for Fundamental Physics: The Astrophysics of Neutrinos, Axions, and Other Weakly Interacting Particles (Chicago, IL: Univ. Chicago Press)

Richard, O., Michaud, G., Richer, J., Turcotte, S., Turck-Chièze, S., & VandenBerg, D. A. 2002, ApJ, 568, 979

Rodrigues, T. S., Bossini, D., Miglio, A., et al. 2017, MNRAS, 467, 1433

Rui, N. Z., & Fuller, J. 2021, MNRAS, 508, 1618

Salaris, M., & Cassisi, S. 2006, Evolution of Stars and Stellar Populations (New York: Wiley)

Schinder, P. J., Schramm, D. N., Wiita, P. J., Margolis, S. H., & Tubbs, D. L. 1987, ApJ, 313, 531

Schönberg, M., & Chandrasekhar, S. 1942, ApJ, 96, 161

Schröder, K. P., & Connon Smith, R. 2008, MNRAS, 386, 155

Schwarzschild, M. 1958, Structure and Evolution of the Stars (Princeton, NJ: Princeton Univ. Press)

Schwarzschild, M., & Härm, R. 1965, ApJ, 142, 855

Scuflaire, R., Théado, S., Montalbán, J., et al. 2008, Ap&SS, 316, 83

Soderblom, D. R. 2010, ARA&A, 48, 581

Sweigart, A. V., & Gross, P. G. 1978, ApJS, 36, 405

Takata, M. 2016, PASJ, 68, 109

Tassoul, M. 1980, ApJS, 43, 469

Théado, S., Alecian, G., LeBlanc, F., & Vauclair, S. 2012, A&A, 546, A100

Thomas, H.-C. 1967, ZA, 67, 420

Thoul, A. A., Bahcall, J. N., & Loeb, A. 1994, ApJ, 421, 828

Ulrich, R. K. 1986, ApJ, 306, L37

Valle, G., Dell'Omodarme, M., Prada Moroni, P. G., & Degl'Innocenti, S. 2016, A&A, 587, A16

van Rossem, W. E., Miglio, A., & Montalbán, J. 2024, A&A, 691, A177

Vauclair, S. 2003, Ap&SS, 284, 205

Ventura, P., D'Antona, F., & Mazzitelli, I. 2008, Ap&SS, 316, 93

Vorontsov, S. V. 1988, Advances in Helio- and Asteroseismology, IAU Symp. 123 ed. J. Christensen-Dalsgaard, & S. Frandsen (Dordrecht: Reidel Publishing) 151

Vrard, M., Cunha, M. S., Bossini, D., Avelino, P. P., Corsaro, E., & Mosser, B. 2022, NatCo, 13, 7553

Willett, E. 2023, PhD thesis, Univ. Birmingham Available at https://etheses.bham.ac.uk//id/eprint/14396/

Yu, J., Huber, D., Bedding, T. R., et al. 2018, ApJS, 236, 42

Zahn, J.-P. 1991, A&A, 252, 179